DCPL0800042532

be returned on or before the last date

already requested by

Burning

D1494244

SANTRY BOOKSTORE
DUBLIN CITY PUBLIC LIBRARIES

3145

Ireland's Burning

How climate change will affect you

PAUL CUNNINGHAM
RTÉ Environment Correspondent

SANTRY BOOKSTORE
DUBLIN CITY PUBLIC LIBRARIES

POOLBEG

Published 2008
by Poolbeg Press Ltd
123 Grange Hill, Baldoyle
Dublin 13, Ireland
E-mail: poolbeg@poolbeg.com
www.poolbeg.com

© Paul Cunningham 2008

Copyright for typesetting, layout, design
© Poolbeg Press Ltd

The moral right of the author has been asserted.

13 5 7 9 10 8 6 4 2

A catalogue record for this book is available from the British Library.

ISBN 978-1-84223-331-3

All rights reserved. No part of this publication may be reproduced or transmitted in any form or by any means, electronic or mechanical, including photography, recording, or any information storage or retrieval system, without permission in writing from the publisher. The book is sold subject to the condition that it shall not, by way of trade or otherwise, be lent, resold or otherwise circulated without the publisher's prior consent in any form of binding or cover other than that in which it is published and without a similar condition, including this condition, being imposed on the subsequent purchaser.

Mixed Sources
Product group from well-managed
forests and other controlled sources
www.fsc.org Cert no. TT-COC- 002341
© 1996 Forest Stewardship Council
FSC

Typeset by Patricia Hope in Sabon 11.5/15.5
Printed in the UK by CPI Mackays, Chatham, ME5 8TD

About the Author

Paul Cunningham has been RTÉ's Environment Correspondent since 2001. In that time, he has reported extensively on the issue of global warming, both domestically and internationally. He has also reported on numerous wars including Bosnia, Algeria and Darfur. Prior to his appointment as correspondent, Paul reported on the infection of Irish haemophiliacs with HIV and hepatitis C from contaminated blood products. For his work, he won the ESB National Radio Journalist of the Year award and an Irish Film and Television Award. His first book, *A Case of Bad Blood*, with Rosemary Daly, was also published by Poolbeg Press.

Terenure Branch Tel: 4907035

Acknowledgements

At the outset, I would like to express my deepest gratitude to all of the people who agreed to be interviewed for and feature in this book – particularly given its personal focus.

I would especially like to thank my editor, Brian Langan of Poolbeg Press. He called me – out of the blue – enquiring if I might have an idea for a book. Craftily, he ensured that five years had passed since I'd written the previous one, with my friend Rosemary Daly. The time-lag meant I couldn't remember precisely how much work was involved. Despite my broken promises and broken deadlines, he quietly applied irresistible pressure which has resulted in *Ireland's Burning* becoming a reality. His advice and insight were immense. All at Poolbeg have been a pleasure to work with.

It simply wouldn't have been possible to write anything unless my employer, RTÉ News, turned a blind eye, from time to time, to my absences. For that, I'm indebted to our Director of News, Ed Mulhall, but also the Duty Editors and Programme Editors who run Ireland's largest newsroom. My parents, Connell and Mary Cunningham, also provided a sanctuary to escape to.

Many people have assisted me in many ways. Thanks to Donal Buckley, Treacy Hogan, Frank McDonald, Mark Little, Tom MacSweeney, Joyce Jackson, Barrie Hanley,

Andrew Parish, Richard Dowling, Elaine Keogh, Deirdre O'Brien, Liam Reid, Olive Stephens, Bríd McGrath, Bill Whelan, Niall Madigan, Nuala Cunningham, Kate Horgan, Stephen Mackin, David McCullagh, Neilus Dennehy, Michael Cassidy, Niamh Leahy, John Lambe, Niall McGuigan, Sarah Malone, Emma McNamara, Caitríona Perry, Piaras Kelly, Niall Hatch, and not forgetting DJ Dirk Montage.

Finally, to Aisling, Isabelle and James – sorry for tap-tap-tapping for so long. It'll be years before it happens again.

Paul Cunningham
April 2008

For Flor

CONTENTS

Introduction

The heavily armed militia ambled their way down the red-grit hill towards our jeep. Four rode on camels, one was on a donkey and two were on foot. They made no attempt to conceal their high-powered semi-automatic weapons. More armed men were undoubtedly following them. This was Darfur, on the rim of the Sahara, and we had no idea what was going to happen next.

RTÉ cameraman Michael Cassidy and I had travelled to the region to try to film this feared force, known as the Janjaweed. These Arab nomads were accused by the United Nations of waging a vicious war against the settled African farmers of the area. With the help of the aid agency GOAL, we saw both the miserable refugee camps where the farmers were living and the burnt-out villages to which they would never return. But no Janjaweed could be found. Now, just as we were taking the long drive to the airport and home, the Janjaweed found us.

One man wearing turquoise fatigues gestured to us that he would like something to drink. I got out of the jeep and brought over a plastic litre bottle of water. The dust swirled around as he grabbed the icy drink, lowered a huge volume and then passed it on to his comrade. Taking my cue from his big thirst-quenched smile, I pointed to my colleague Michael who had raised his camera. My new-found Janjaweed militia friend smiled again. We were in. Few TV crews in the world would get so close.

As Michael began to film, I watched with amazement as more than two thousand camels lurched past us. All probably war booty. Every now and again, more militia members would come into view. One had a special satellite phone – a hand-held unit which was more up-to-date than mine. To me, it was further confirmation that the Janjaweed were working in collaboration with the Sudanese army.

A few smiled our way but most of the fifty or so militia members either ignored us or fixed us with menacing stares. One man in olive fatigues held my gaze for what seemed an age, his hand remaining on his shiny gun throughout. Michael took his shots quickly and we returned to the jeep, not forgetting to hide the tape so that if the camera was taken, we would still have the shots.

After they were gone, euphoria and relief gripped us. We grinned widely as we continued our bumpy journey through desert to the airstrip. We had got the shots and not got shot! Finally things were going our way. That's when our jeep suddenly gave up. As the driver worked on the engine, all we could do was wait.

I stood in the shadow cast over the sand by the jeep and looked at my boots. On television, the desert can seem like a place without life. But, standing in the shade, there were bugs and critters of all shapes and sizes moving about. It reminded me that the Sahara is alive too and expanding all the time. Indeed, the desert is something of a voracious beast. The world's climate is changing and, in Darfur, it has led to reduced rainfall and, in turn, less and less pasture.

This was a big factor in the outbreak of war. The Janjaweed militia come from Sudan's nomadic Arab community, who herd their animals around Darfur. As the rains reduced, they increasingly came into conflict with the landed African farmers over access to water and herding rights. The farmers already felt the Arab-dominated government in Khartoum had abandoned them. Now their lives were becoming intolerable. As this resentment deepened, some began an armed separatist campaign. The Khartoum government's response? They armed the Janjaweed.

The environmental problem certainly wasn't the sole reason for war in Darfur but, as RTÉ's Environment Correspondent, it struck me how this issue had significantly contributed to what was described by the UN in 2004 as the greatest humanitarian disaster in the world. Indeed, according to a report from the UN's Environment Programme in 2007: "Sudan is unlikely to see lasting peace unless widespread and rapidly accelerating environmental degradation is urgently addressed." As I kicked the red sand of the Sahara with my boot, it occured to me that environmental problems around the world – such as increasing desertification – are only going to get worse as a result of global warming. Harsher climates will mean fewer resources and probably more conflicts. I resolved that, if Michael and I ever got out of the desert, I would look into this further.

Now, in 2008, global warming is being described as the defining challenge of our generation. The reason it's given this label is the firm belief that failure to tackle the problem now will result in future generations being afflicted. Yet just two years ago, few paid much attention to climate change which, in turn, probably explains why there is so much confusion over the subject.

What might surprise some people is that global warming is already upon us. Birds are a good indicator of this change, because in general they can move quickly in response to any changes in their habitats. Over the past few years, Ireland has had an influx of Little Egrets – a snow-white member of the heron family normally found in the Mediterranean. A major rarity just a decade ago, they now arrive for winter and breed here. Another recent colonist is the Mediterranean gull.

Birdwatch Ireland is also noticing that some native birds are choosing not to migrate. For example, the chiffchaff, a common warbler, was until recently strictly a summer resident. But over the past decade, an increasing proportion are opting not to depart for southern Europe and North Africa each autumn.

But if the pace of change is rapid, that can cause further problems. That's because birds are very susceptible and vulnerable to any change in conditions during their breeding cycle. Take the blue tit as an example. While the adults eat a wide range of foods, its chicks demand one type of food more than any other – protein- and mineral-rich caterpillars. As a result, their hatching is timed to coincide with the emergence of the caterpillars. However, it's thought that warmer temperatures could trigger the chicks to emerge earlier than usual. With no caterpillars and subsequent cold-snaps, the survival of the blue tit could be endangered. According to the best predictions of the United Nations, one-third of all species must either adapt or face extinction.

In Ireland we tend to comfort ourselves that we won't face catastrophic consequences here, but the impacts could well be severe – one-fifth of native plant species could be gone by 2050; flooding is likely to increase significantly, particularly in the Shannon region; coastal erosion is projected to become far more severe; Dublin, with its expanded population, will run out of water from current sources. The list goes on.

Why is it happening? How quickly will it happen? How varied will the consequences be? What, if anything, can we do about it?

Just as there are so many questions surrounding the subject, there is no such thing as *the* answer to climate change. So the aim of this book is to explain the scale of the problem, and the complexity involved, through the direct experience of a range of people. Over the following twenty-one chapters, you will learn about the lives and views of politicians, scientists, entrepreneurs, commuters, farmers, inventors, sceptics, schoolchildren and green campaigners. Their diverse views, which sometimes clash, will inform you about the extensive range of climate change impacts facing this country; what solutions are on offer; how previous opportunities to tackle the problem have been missed; and, critically, the difficulties of striking a balance between the needs of living today and the obligation to the generations of tomorrow.

The book will simplify the complex issues around climate change, without ever being simplistic. Hopefully, it will also play a role in dispelling the idea that climate change is merely an "environmental" problem instead of something which permeates every aspect of our lives.

According to the government's latest information campaign, global warming is the single greatest challenge facing the planet and the way we tackle the problem will define our generation. If the UN's scientists are to be believed, we've got less than two decades to shoulder the burden and make substantial progress. Otherwise our planet, and future generations who live on it, are going to be in deep, deep trouble.

PART ONE

What do we know?

"As we know, there are known knowns; there are things we know we know. We also know there are known unknowns; that is to say, we know there are some things we do not know. But there are also unknown unknowns -- the ones we don't know we don't know."

-- Donald Rumsfeld, *Former US Defense Secretary*

1

Russian Roulette

Ireland's most famous TV weather forecaster is unequivocal – science has proven that the planet is heating up as a result of human activity. He believes the consequences could be negative in the extreme, unless we act immediately. Maybe it's not surprising that he is a worried man. The public simply does not seem to care. "I get worried, actually, by what little understanding people have of it; by how many people still think that it's just a type of crackpot scientific theory that will be replaced by something else next week."

Gerald Fleming speaks to me with a speed and passion which is slightly unexpected. Like most TV viewers, I'm used to hearing him delivering his forecasts in a soft Wexford tone – finishing off with a friendly smile and trademark wink. Clearly, climate change is no joking matter. "I really think that people in Ireland don't understand how profound this issue is. They think of environmental issues as the preserve of sandal-wearing weirdos, but we all live in the environment."

A recent trip to the Young Scientist Exhibition at the RDS in Dublin clearly hadn't helped matters. He had been attracted to one project which was entitled: 'Does climate change affect where I live?' The students who drafted the project came from Kerry, but when Gerald asked them what they had found, the response was: "We don't have much pollution in Kerry." The inaccurate linkage between local pollution and the global problem of climate change sets him off. "I don't expect people to understand the science, but it's a global problem we as humanity have to face up to." I ask him why he thinks the penny hasn't dropped yet. He muses: "Maybe it's because the economy has been strong here for the last ten years, and we feel we can buy our way out of trouble."

Part of the reason Gerald Fleming is exasperated is because he's been tracking this phenomenon for twenty years. Global warming may well have become a regular topic of conversation over the past eighteen months, but it's "old hat" to those working in the field of weather. "Certainly it was coming into the literature, the normal stuff that we read, by the mid-to-late 1980s. The science wasn't that far advanced, but it was becoming an issue. We knew it was going to come to the forefront."

The mid-1980s was also when Gerald's TV career took off. He presented his first television forecast in July 1985 and hasn't looked back. He's now co-ordinator or leader of the weather team in RTÉ. It might be surprising to some, but he views his job as being far more than simply predicting if it's going to be sunny in Donegal or raining in Kilkenny. A core component is also explaining what's happening in the weather system, and the projected impact of climate change. He hopes that he and his colleagues can have an impact on the public's perception. "As a weather forecaster, I hope people have some sort of sense that I have some credibility in this business. All of us feel a tremendous responsibility to use what profile we have and what credibility we've built up. The challenge for me and my colleagues is to present the science, explain it, and say to people: 'This is something you are going to have to take seriously.'"

The RTÉ weather team is constantly analysing how it can improve communications. Gerald started in the days when a weather forecast was delivered by placing magnetic clouds onto a cardboard map of the country and pointing to regions with a long thin wooden stick. While computer wizardry makes everything look so exciting now, it's a technology that is also being used to inform and educate. For example, at the end of 2007, Evelyn Cusack used part of the weather bulletin to show how Met Éireann statistics proved the year gone by had been one of the warmest on record.

Given the torrential summer downpours, the data for 2007 was perplexing to most people. How could the year be classified as warm when the summer was a washout? For Gerald, it's easy to explain: "People equate warm weather with sunshine. But warm weather often comes with rain in this country. The warmer air is, the more moisture it can hold. There's a very strong relationship between the temperature of air and the water it can carry. We had a lot of very mild air over the country last summer and it was carrying an awful lot of moisture." It seems when you throw in warmer night temperatures, and a few hot weeks in April and September, you end up with one of the hottest years on record.

This link between the weather we're experiencing and future impacts of climate change is going to become far more common. RTÉ's *Morning Ireland* radio programme has asked the forecasters to analyse the latest climatic data on a monthly basis. It's a request which Gerald is delighted to fulfil. "The more information people have, the more they are able to weigh it up. The impacts on society of global warming are going to be so great that we are going to have to take people with us. If people don't understand that there's a reason why I can't drive my car every day, there's a reason why I can't fill my tank full of oil, it could be dodgy. It's about trying to raise awareness that there are tough decisions to take."

As there is so much talk about global warming, let's go back to basics. The earth receives energy from the sun. Thirty per cent of it is bounced back into space, but the bulk of it reaches the surface of our planet. The earth in turn begins to emit its own heat, some of which escapes into space but, critically, the rest is absorbed by the greenhouse gases in our atmosphere. These gases are vital to our very existence because they function as a blanket keeping the planet warm. Without this layer, the earth would be colder by more than thirty degrees Celsius.

There are many greenhouse gases – the most well-known are carbon dioxide, methane and nitrous oxide. Then there are the unpronounceable ones such as Perfluorethane, Perfluorocyclobutane and, the one that trips off the tongue, Trifluoroiodomethane. Water vapour is also very important as it too retains heat.

The problem is that human activity has, since industrial times, dramatically increased the amount of greenhouse gases in the atmosphere. According to the UN, eighty per cent of man-made carbon dioxide currently comes from the burning of oil, coal and gas while the remainder comes from deforestation and land-use changes. The overwhelming view of science is that by increasing greenhouse gases, we're effectively turning up the heat on Planet Earth – and that is potentially catastrophic.

Given his clear commitment to the issue, it may come as something of a surprise that Gerald Fleming wasn't a weather buff when young. "I wasn't a weather anorak. There are people who really live everything to do with the weather – that was never me. I studied science in the local Christian Brothers secondary school and, with very good teachers, ended up studying physics at university. My masters degree happened to be in the atmospheric physics area and, when I came out of college in the late 1970s, jobs were not thick on the ground." With that background, you either went into the old Department of Posts and Telegraphs or the Met Service. Gerald chose Met Éireann.

After just under a year-and-a-half of training, he was stationed at Dublin Airport and provided forecasts to pilots. Within two years, however, he was posted to the Central Analysis and Forecast Office in Glasnevin, with its glass pyramid-style design. It was here that he began to pick up on the theory we know today as global warming. His physics background was a trigger. "It would have been of interest. I could understand the physics behind climate change, about why gases are radiantly active. It wasn't a big difficulty."

Thinking back on the problems faced by his predecessors working on the RTÉ broadcasts, it still fills him with awe that they got anything on air. "Our knowledge of how the weather behaves, how the atmosphere changes, has improved hugely in the past twenty-five years. When I think of people like Paddy McHugh and Charlie Daly, who were forecasting back in the 1960s and 1970s, I mean it really must have been on a wing-and-a-prayer. They had experience to go on – but they had little else: a few charts, a twenty-four-hour forecast and that was it."

In 2008 it's very different, with super-computers available to crunch the data from the orbiting satellites and multiple weather stations. "Now the mathematical models of the atmosphere have become much more accurate in plotting its current state, and moving it forward in time. That's mainly through scientific instrumentation and more satellites. We know a lot more about the atmosphere and we know more about how it behaves."

Yet he also accepts that forecasting has its perils, despite the technology. "The atmosphere is inherently a chaotic system. You are always going to be faced with uncertainties." That said, it used to be believed that you couldn't get a useful forecast beyond a week due to the chaos. Now lengthy forecasting can give projections over several months in certain parts of the world.

One secret which Gerald lets slip is that the accuracy of a forecast can partly depend on where the forecaster hails from. "If you have an understanding of the location

– of the hills and coastline – it can help you refine the forecast. I'm from Wexford myself and would know the area. I would probably give a far more accurate forecast from Wexford than from Mayo, which is a place I don't know all that well." The computers are continually improving and providing better regional projections but they are still a long way off from giving detailed, localised weather information.

However, the reliance and trust being placed on computers is of key importance in the context of climate change. That's because the software used to provide short-term weather forecasts is very similar to that deployed to give predictions on global warming impacts. For Gerald, the public can trust what the machines are telling us. "We're very familiar with them. They are not perfect – we understand their flaws and allow for them. But we also know they are far, far better than they were ten years ago and are constantly improving. They represent our most complete understanding of the atmosphere."

Despite Gerald's firm views, which are clearly based on long experience, the public has some way to go in fully accepting that computer models can give an accurate picture of the impacts of climate change on Ireland in fifty years' time. It may be interesting to people on the street, but I think it's fair to say it does not inform their decision-making too deeply. For now, at least.

Yet Gerald Fleming is by no means sanctimonious about this issue. He and his wife, architect Mary Duggan, have three children and face the tension between high principle and street practicalities like everyone else. "We have certainly worked on our house to make it more energy-efficient – putting in solar panels and improving the insulation. Having said that, I'm also trying to run my life and know how difficult it is to fit these concerns into a daily reality." And he's a busy man. Apart from the family and day job, he also has a keen interest in the arts, serving for many years as Chair of the Wexford Arts Centre. More recently he's been the Chair of Wexford Swimming Club and, somehow, manages to shoe-horn in a regular appearance at the Wexford Wine Circle.

As well as being active, he's also honest. The day we meet in Glasnevin, there had been several downpours and I am interested in what mode of transport he uses to get to Met Éireann headquarters. Gerald often cycles but, that morning at least, he was thinking of creature-comforts rather than saving the planet. "It was lashing rain and I was looking out thinking to myself, 'Not on the bicycle.' I drove into work through the middle of Dublin morning rush-hour traffic. It took me an hour to get in. So I was part of the problem as much as anybody." While we're being honest, I had also driven to Glasnevin, rather than taking the bus.

Gerald recognises that awareness of global warming has increased considerably recently and people are taking action. He's just not convinced there is enough momentum. "I think the Power of One campaign is very good and very useful. However, an individual can only do so much. At the top level, you need government to be doing things. For example, you can't use public transport unless it's there. The government is, to an extent, kicking in now, but I'm not sure about local authorities or town councils."

One optimistic sign has been the recent government effort to link a financial penalty with the production of pollution. "One indication of change is the tax regime for cars, which is now going to be based on carbon dioxide emissions rather than the size of the engine. When this gets up and running there will be winners and losers. It will give us a clear indication of how much we've progressed, when it's possible to examine how people spend their money when purchasing cars – how many four-by-fours are sold; how many cars with two-litre engines."

You don't have to be in his company for too long before you understand that Gerald isn't simply informed by the public debate in Ireland alone. As well as reading the literature, he's been a participant in the wider international analysis. "My expertise is in the quasi-scientific area of science communication. We have a working group in the World Meteorological Organisation, or WMO, which aims to help Met Services to improve the presentation of their product to their community. Because Met Services are organisations of scientists, very often they don't have in-house experience in media and communications. Clearly they need help to make the best use of what they have to offer society."

Gerald's expertise is recognised internationally and reflected in his curriculum vitae: he's a two-term chair of the International Association of Broadcast Meteorology; the co-chair of the first world conference on broadcast meteorology in Barcelona in 2004; chair of the expert team on media issues for the WMO; co-chair of the media committee of the European Meteorological Society. When not focusing on that, he has been giving talks and seminars at the UK Royal Met Society, the American Met Society and the BBC. He is, as would be expected, also a valued member of the Irish Meteorological Society.

Met Éireann is affiliated to the WMO, which was significantly involved in forming a scientific advisory committee for the United Nations on global warming. The organisation would end up being called the Intergovernmental Panel on Climate Change and was co-awarded the Nobel Peace Prize in 2007, along with Al Gore. The IPCC's aim was to review data from around the world on evidence of climate change and then issue a report every five years or so to assist and influence international government policy.

In 2007, the IPCC issued a series of four reports. The first came from its physical scientists – experts on what's currently happening to the planet. They concluded that "widespread changes in extreme temperatures have been observed over the last fifty years; heat waves have become more frequent; more intense and longer droughts have been observed over wider areas since the 1970s; the frequency of heavy precipitation events has increased over most land areas; there is observational evidence for an increase of intense tropical cyclone activity in the North Atlantic."

Gerald was following the IPCC closely from its inception. By the time the work for the Third Assessment Report was underway in the 1990s, he felt there was no scientific doubt remaining that human activity was responsible for global warming. "It was very clear, at that stage, the scientific consensus was there. Yet science is such that you always have to try to keep an open mind. This is where science and media often collide. If a media organisation is dealing with any topic, they like to bring people from different sides into the story. But on this topic, you have ninety-nine per cent of scientists backing the IPCC findings and one per cent questioning them. If you are reading stuff in the paper, you might get the idea that there are two equal opinions, when the equality isn't there. That was the situation we were dealing with a lot in the 1990s."

On the other hand, the IPCC itself is not beyond criticism. The way the organisation functions is that more than 2,000 scientists and reviewers evaluate publications on climate change and, after several years' work, complete a lengthy report and summary. But the document still isn't published at that stage. Instead, the material is handed over to government representatives from more than a hundred countries who edit the summary. To its detractors, such a political involvement is unscientific and devalues the report. To IPCC supporters, the political involvement is essential as, once agreed, the world's governments are duty-bound to act.

This old argument re-emerged in 2007, when the IPCC published its Fourth Assessment Report. The executive summary of the work of the physical scientists concluded that the IPCC was now more than ninety per cent certain that human activity was responsible for global warming. Because that appears to suggest there is nearly a one-in-ten chance of something else being the cause, I ask Gerald if it made him re-think. "I would guess that if the physical scientists were left to themselves, it would have been ninety-nine per cent. Having said that, no good scientist will ever completely rubbish the other point of view because we have to try and keep an open mind. That's difficult for the lay-person to understand –

keeping an open mind is part of science, but it does not mean you doubt your findings in any real sense."

While the IPCC said its forecasts were based on the best evidence, it was not in a position to factor in things like what was happening in Antarctica. The reason? The data wasn't extensive or solid enough. Do these gaps of knowledge lead Gerald to question that humanity is responsible for climate change? "No. Not to my mind. The evidence is there in terms of the rise in carbon dioxide. It's clear that's come because we're burning fossil fuels. There's no other plausible explanation."

After the analysis of what the IPCC had to say, some commentators suggested that it was the sun, rather than mankind, which was responsible. Gerald dismisses the suggestion out of hand. "The energy we get from the sun is changing constantly. People have a fairly good understanding now of things like the sun-spot cycles. But no matter what the sun does, the measurement of carbon dioxide in the atmosphere is very solid. If it's not Leaving Cert physics, it's first or second year college physics. In fact, one of the first people to realise that this may have been a problem was John Tyndall from Carlow at the end of the nineteenth century. The rise in temperatures is not a surprise."

In the main, Met Éireann focuses on collecting hard data on the situation in Ireland. Worryingly, there are already trends which are being linked to global warming. "As a forecaster, and sitting at the forecast bench, the one thing which jumps out at us is the decrease in the number of frost nights at wintertime. It's the winter nighttime temperatures in particular that seem to have changed more quickly than the other parameters. The weather always has changeability built into it, so you can only interpret these things over long-scale periods. But in milder winter nights, I think I can see a significant change from when I started forecasting. That's not today or yesterday, but it's not a hundred years ago either."

I press him on these trends but he is reluctant to talk about anecdotal evidence when climate change needs to be considered over long periods of time. But after a bit of pushing, he volunteers one observation: "We used to have frosty nights. We used to have cold nights. But how many kids born since 1990 have seen significant snowfalls in this county? Not very many. Whereas those of us born in the 1950s, 1960s and 1970s remember significant snowfalls when we were younger."

The idea that the public is unaware of the climate change beast which is bearing down on them is a theme which Gerald returns to again and again. He quotes a recently deceased expert on climate change who contended that the basic scientific information about global warming was known fifteen years ago – it just wasn't communicated adequately. And every year that passes, with more CO_2 pumped into the atmosphere, is not just a year lost, but an addition to the problem.

It's a view which informs, if not drives Gerald on. And this is because, according to the UN and EU, dangerous or irreversible climate change could happen quite quickly. They estimate that when the world heats up by two degrees above pre-industrial levels, many parts of the planet's eco-system will simply not be able to cope anymore. According to Gerald, we've already increased by 0.7 of a degree. He

says temperatures will increase by a further 0.7 degree –
no matter what we do – due to emissions which have
already taken place. In his words, there's not much head-
room left.

When he spells out the statistics, it seems to me that
he is painting a picture in which the people of Ireland,
and the people of the world, are playing Russian roulette,
but don't even know they have a gun in their hand. What
does he make of the analogy? "It's a good way of putting
it. We could happily sail along for the next ten years and
suddenly find that, oops, the limit is going to be breached.
And at that point, we may have passed the point of no
return. That's the challenge – to get people to understand
that."

2

Noah's Ark

Saving Ireland's native plant species, you might be surprised to learn, is work often undertaken in great secrecy. According to Peter Wyse Jackson, Director of the National Botanic Gardens, nowhere more so than in Killarney, County Kerry. "The endangered Killarney fern is a fascinating story. It's a very delicate fern with translucent green leaves just about one cell thick. If it dries out, it dies." The fern used to be common and there are many accounts of the lakes in and around Killarney shimmering with this translucent fern. Not any more. "During the Victorian period there was a fern craze, where ferns were collected and sold to tourists. Essentially the plant was wiped out in Ireland from over-collecting." What's exciting botanists now is that the plant is back. "There are specimens in the wild – it's hard to say how many, probably fewer than twenty clumps – hiding in the rocks near waterfalls."

The Killarney fern suffers a particular problem, due to its unique lifecycle. "It lives in two forms – a sort of leafy frond which then produces spores that grow into a green algae, a little thread-like plant. For some reason, it never again develops into the leafy shoots. So we are doing research on that – to see if we can find out how to develop it and re-establish the fern."

However, because it has such a cachet, the location of the twenty remaining clumps has not been disclosed to the public. "Many of them are known by botanists but the information is kept secret, as someone would pull it up again and go off with it. In the past, the localities of some of these really special plants were handed down from elderly botanists to the next generation. It was a rite of passage."

Global warming, however, is the biggest threat to our native plants. Peter has predicted – in his words "conservatively" – that twenty per cent of the Irish wild plant population appears to be particularly vulnerable to climate change. In other words, 170 species here could be wiped out by 2050 due to changes in temperature and habitats. It's unheard of for a story on native plants to make headline news. Yet Peter's research paper was considered of such importance that Ireland's flora was catapulted to the top of the main evening news on RTÉ TV in late 2007.

The research paper is still on the website of the National Botanic Gardens and gives some insight into what motivates the Director. He notes a statement by the former Taoiseach, Bertie Ahern, in the National Climate Change Strategy: "Climate change is among the greatest challenges of our time." But Peter goes on, somewhat acidly, to observe: ". . . the Strategy itself pays little attention to the impact of climate change on Ireland's biodiversity, but instead concentrates on issues such as energy supplies, transport, industry, agriculture and waste." For Peter Wyse Jackson, a big alarm bell has gone off, but he's unsure if either the government or the public has heard it.

Sitting in his office overlooking the splendid gardens in Glasnevin, Peter tells me that he always had a keen interest in plants. "It irritated me that I couldn't identify them. I like to know what is around me. As a student in secondary school, at St Columba's in Rathfarnham, I would bring in plants to my biology teacher, Richard McMullen, and say 'What is this?' He generally knew. He suggested I buy a book called *An Irish Flora* by D.A. Webb, which had keys, so you could sit down, take out your plant and work it out." During that summer, he tried to learn how to identify a new plant every day. The seed was planted.

There was an intellectual rather than emotional drive to his studies. "What I find fascinating are the stories behind plants. If I look at a daisy, I don't find it intrinsically fascinating. Some people do. Some people just find plants wonderful in all their textures. It's the significance, and the stories associated with them, which really appeal to me."

Given the personal course of study he had taken in school, it wasn't much of a surprise when he went on to take up Botany at Trinity College Dublin. After completing a masters and PhD, he secured the plum job of Curator of TCD's Botanic Gardens in Dartry, which were first established in 1687. The trigger to get involved in plant conservation came in 1985 when he was lucky enough to secure an expedition to Mauritius. "I worked with a lot of rare and endangered species. I was amazed to find a very rare hibiscus, which I collected and propagated and doubled the world population. There were only two specimens known and I grew two more. And I thought to myself, *Goodness, there's an awful lot to be done in conservation!*"

Two years later, Peter took a year-long sabbatical to help set up a new world network for botanic gardens. He ended up staying for eighteen years, twelve as head of the organisation. When Peter started, there were twelve members. Today it has more than 700. That growth is part of an awakening about the value of plant conservation and the important role in that campaign played by botanic gardens. One hundred and twenty countries around the world now have botanic gardens – fifty-four per cent of them have been created in the past fifty years. But when the job of Director of Ireland's National Botanic Gardens came up three years ago, Peter Wyse Jackson felt it was time to come home. Not surprisingly, conserving Ireland's native flora is now a cornerstone of his work.

Given the extensive coverage he received for his paper on endangered Irish flora, I ask Peter to outline how climate change can impact on plant life. "Climate has a huge impact on plant distributions. The patterns of plant distributions around the world has been based on both present-day and former climatic patterns. So, for example, we know that during the last ice period in Europe, a lot of the plant species moved south with the advancing ice. When the ice retreated, they headed back north again."

This concept of natural migration, as a response to changes in temperature, takes a little bit of time to understand. How do plants *move*? "Well, they are not like birds; they don't simply say, 'It's getting too cold', and head off. But what happens is, plants have certain climatic tolerances; they will grow where the climate and conditions are best suited to them. If the conditions change farther north or farther south, they will gradually, through seed dispersal, get established in those regions. When you look back, after the ice-age in Ireland, an oak would move about a mile a year on average. A lot of the mechanisms for seed dispersal are so varied. Squirrels, for example, will take acorns and bury them. The rate of dispersal will therefore be related to the distances that squirrels move."

Plants are clearly robust and can react to changes in their habitat. However, global warming presents a far bigger problem, and for many plants an insurmountable one. "If climate change was happening as a slow process, then the plants would be able to move to adapt to the changed conditions. The big problem is that it's happening very quickly."

This scenario is further complicated by the construction boom which has resulted in plants being incapable of moving as they once did. This, Peter maintains, is a hugely important factor when assessing the ability of a plant to survive the consequences of climate change. "Most of them have nowhere to go to. So many species occur in habitats that have restricted distribution. If you had a bogland species, in the past it could hop from one bog to the next. But we're now left with just a handful of suitable habitats. So they are land-locked, essentially. And that's why, if we lose plants in Ireland, it's because their distribution is already so whittled down in any case." While we may cordon off certain areas, it does not necessarily mean that the species inside them will survive either. In Peter's view, it's a bit like protecting the stable when the horse is dead inside. One possible solution is to expand our hedgerows and ensure that native species are deployed there.

What made Peter's research paper stand out was the fact that he listed all 170 species under threat. Those on the list included the three-lobed crowfoot, shepherd's cress, sea-kale, mountain pansy, eight-stamened waterwort, rough clover, hairy bird's-foot-trefoil, sea pea and meadow saxifrage. But given that the science of climate change is usually qualified, I ask him to explain how the list was compiled. "I ran it through a filter of what plants we believe are particularly vulnerable to climate change. For example, those that have nowhere to go; those that have particularly low habitat tolerances; those that are of restricted distribution; those that may be particularly associated with or reliant on other species – for example, if there are no squirrels, there is no distribution for acorns. I also then looked at the predicted changes in the climate. If we assume a 2.5-degree change in July temperatures by 2050, then the species that are hanging on the northern coast, because they are already at the limit of their climatic conditions, are simply going to drop off the cliffs."

The difficulty with climate change is how to produce hard data which you can stand over. Inevitably, there are qualifications. So given his scientific background, how comfortable was Peter with his own work? "I was not really happy, but we've nothing else. We can't say, about climate change, that in January you will have this impact and in February you will have this impact. So climate change itself is a pretty imprecise science. A lot of it, like the predicted extreme weather events . . . you don't know when it's going to come. We can't be more precise, but we just have to make a best guess."

That said, the trend of increasingly warmer night temperatures is clear from our weather stations. According to Peter, that's also going to have an impact. "If night temperatures have gone up, it will allow more organisms to survive in Ireland that might not have otherwise. We all know that if we put out our sweet-pea too early and there's a late frost, it will get killed. I suspect the impact will be greatest on what species can grow and survive in Ireland, rather than what we lose as a result of it."

One plant, the Cornish moneywort, is already under pressure. Once fairly common on the Dingle peninsula, it is now becoming more difficult to find. "It was fairly abundant on the north slopes of the peninsula, on wet banks and bases of walls. That's becoming increasingly rare and endangered. In my lifetime, there are lots of habitats where it no longer occurs. The plant you do find in those particular habitats is the New Zealand willow herb, which spread as a garden plant and now is invasive. It seems to be ousting the Cornish moneywort. As the climate improves, more and more plants may escape. We will probably also see a lot of pests and diseases that are more prevalent in northern Europe attacking native plants."

The key to understanding Peter Wyse Jackson is the value he places on native plants. "Ireland has a fairly small flora – it has just over 900 native plants. Those are ones which were not introduced through any human agency. Some of them we just can't tell; there are perhaps fifty species and we just don't know if they are native or not. They might have been introduced as weeds of crops by the first farmers. The ones I'm really interested in are the native."

But does the replacement of one plant, albeit native, by another amount to a problem? Have we necessarily anything to worry about? For Peter Wyse Jackson, there is no equivocation. "You could say it does not matter, but they are a part of Ireland's heritage, our natural environment. They are every bit as important in Ireland as the paintings in the National Gallery. If we were to go into the National Gallery and remove twenty per cent of the paintings and burn them – even though most people in Ireland may not have seen that twenty per cent, they would still get angry about it. It should be the same with the native flora. It's part of who we are."

In the Ireland of the twenty-first century, the sentiment that native flora is part of our identity would probably seem odd to a lot of people. Peter accepts that common knowledge of the identity and value of our plants is quite limited. "At a general level, people are not aware of what is a native plant. I mean, most people are surprised when you tell them that fuchsia comes from southern Chile. And monbretia in the hedgerows is essentially a garden-escapee and a hybrid from two South African species. I think there is a general disconnection between how people appreciated plants in the past and how they do now. If you were to ask people how they use wild plants around the countryside, people would say they rub dock leaves on a nettle sting. Maybe as children they've played "clocks" – blowing dandelion seeds to guess the time. The hundreds of uses known about in the past have gone. People have forgotten. But there is a great, growing interest in the rediscovery of this."

That interest has spurred on Peter Wyse Jackson to undertake a study of what Ireland's native plants were used for. Clearly the aim is to rekindle interest in the flora and hopefully reinstate their value in the public consciousness. "If you go back in time you would find that native flora was probably the most important resource that people had. I've done work on the uses of Irish plants and found that over 200 were used as medicine because there was no other medical care. More than 100 were used for food. There were flora used for fibres. About 300 to 400 plants had a documented use for rural life in Ireland."

He walks over to his desk and returns with a weighty manuscript which he hopes will ultimately be published as a book. Leafing through the pages, he stops at the medicine section and identifies some of the 240 plants which were once commonly used to treat ailments. "Hemlock leaves were used in combination with linseed meal as a pain reliever. It was regarded in Belleek, on the Donegal/Fermanagh border, as being good for scrapes and bruises. You would mash up the leaves and put it on joints to treat rheumatism, swellings and sores. In the Aran Islands it was put on to treat an abscess." Another plant which grows in meadows, the eyebright, was mashed into solution and, funnily enough, was used to treat eye problems. Poisons derived from plants were also used, including one used in fishing. "The Irish spurge was used as a poison. It only occurs in Cork and Kerry. It would be cut and put into sacks and dumped in the river. It would kill the fish, which could then be harvested."

My eye was drawn to the heading "Aphrodisiacs". Apparently, ten plants with qualities that reputedly enhance libido are growing in Ireland. The Director becomes somewhat coy at my interest. "I can't say if they actually work or not. They are documented, but I'm not sure if I should go into them." It seems that while some of the plants were used as aphrodisiacs elsewhere, it's unclear if they were used here. Why? "People are not going to admit very often if it was used, and I'm not sure I want to promote it!" One plant used by the Navaho Indians as an aphrodisiac is widespread in Ireland, but there are other examples. "There's a spurge in the Isle of Man, one of whose common names is Saturday Night Pepper because it produces a white latex which can be rubbed on to enhance sexual pleasure." For those who are interested, nettles are apparently good for flagellation.

With the onset of climate change, there's now an urgency to protect our native species. The National Botanic Garden is going to be playing a key role in that campaign. "We have a great role in education, where the gardens are being seen as a centre for environmental education for plants in Ireland. One of the ways we can do that is by growing a good variety of Irish plants, so that people see them and appreciate them. Equally we can do specific conservation and research work on plants that are critically endangered. We can both grow them here as back-up collections – in effect, the Noah's Ark for endangered Irish plants – and also work with people managing plants in the wild. We have committed ourselves to having a conservation programme for every critically endangered Irish plant by 2010."

Conservation work is already underway because some plants, such as the Irish fleabane, are in dire trouble. The plant can be found only on the County Tipperary shores of Lough Derg, Ireland's third-largest lake. "It's not found in Britain or western Europe. The nearest place you can find it is Poland, and then it occurs in Siberia. It once was pretty abundant around Lough Derg. Now there's one clump left. We think it's lost out because of the changing water levels caused by the Ardnacrusha power station. It does not produce any seed. What we can do is maintain a back-up collection."

One of the reasons that the Gardens are able to mount a campaign is because the government has begun to loosen the purse strings. "We are gradually putting in place the resources and the people. In the past year, we have doubled the number of research staff here to eight. We have four new research staff, all working on conservation. We are also putting in place new technologies; for example, we hope in the next year to have a DNA sequencer so we can look at the genetics of native Irish plants. If we want to conserve particular plants, we can go and take cuttings of them. But unless we can look at them from a genetic point of view, we're not sure if we are working on identically the same plant."

For many plants, though, it's going to be a struggle, because the process of saving them is complicated. This I learned from a story about a plant in Sligo. "It's a little high mountain plant which has survived in Ireland since the ice retreated. It's got a little rosette at the base of the leaves and a small spike about four inches long and small white flowers. There are only three specimens left on the cliffs on Ben Bulben. It's already endangered and climate change is just going to push it over the edge. So it's become a top priority for us to work on. But it's so rare that it's very difficult to save."

My initial response is to ask, "Why not dig up one of them, to ensure the survival of the species?" But, it appears, that's simply not done. "The risks to the plants in the Botanic Gardens are every bit as great as in the wild. It's well known that if you go into an intensive care unit in hospital you may come out with a disease you didn't have when you went in. It's the same with cultivation. Any rescue operation can end up damaging the plants. So what we have to do is to develop a conservation programme which does not end up diminishing the chances of the plant surviving in the wild." The hope is that science will be able to provide some of the answers. "We have recently developed a tissue culture unit. That is to take small amounts of plant material and to put it into what is essentially a nutrient soup and stimulate the cells to divide. And from those we can do micro-propagation – make plants grow from very tiny bits of material."

One of the factors behind Peter Wyse Jackson's work is that his Irish responsibilities and international interest are about to unite. In June 2010, Glasnevin is going to host possibly the largest ever international conference of Botanic Gardens, involving more than 800 delegates. It will try to ensure that the 1992 UN Convention on Biological Diversity was not just a talking-shop, but a successful strategy, with targets and times, to protect native species. While conservation programmes around the world are making a difference, the sobering reality is that more than 100,000 of the planet's plant species are currently threatened with extinction. And climate change is only going to intensify that threat.

3

Invaders from the Deep

It has been argued that it's going to take a big surprise, if not a big shock, before people in Ireland take climate change seriously. Maybe, just maybe, that shock could come from the sea. Some have speculated that as our coastal waters get warmer, predators might decide to take up residence here. That includes the biggest and scariest predator of all: the great white shark.

Sitting in the Oceanworld Aquarium in Dingle, County Kerry, I look out to sea but can't spot any dorsal fins approaching the harbour at a rate of knots. While the aquarium does have black-tip and white-tip reef sharks, its founder, Kevin Flannery, says a shark the size of "Jaws" will not be appearing off the coast of Ireland any time soon.

But just as I relax, he delivers an unnerving qualification. "There was a very specific case off the Norfolk coast last January. A seal washed ashore leading scientists to conclude:

'Yes, it was bitten by a great white.' Scientists from all over the world have looked at the pictures. The problem is that the evidence is gone; the seal was taken and buried. If we had the seal, and carried out DNA tests on it, we would have been able to say quite categorically: 'Yes – this was a great white.'" So maybe it's not totally safe to go back into the water after all.

What Kevin Flannery says, counts. He's had a life-long interest in the sea and what lives in the salty water. It was an interest first developed by looking into a biscuit tin on a bar counter in Dingle. The publican, Michael Long, had a deep love for natural science. He had a standing agreement with the local fishermen – if you find something of interest, bring it into the bar and you'll get a free pint of beer or stout.

For Kevin, things really took off when he came across an example of a hermit crab, which does not have a shell. This one had apparently fused with a sea anemone, with its round brightly coloured body and stinging tentacles. As with other unusual finds, the quarry was duly brought to Long's bar and placed in the biscuit tin for observation.

Kevin tells me: "It was my first time learning about symbiosis. This anemone lives on the soft back of the hermit crab and protects it. But when the hermit crab is eating, the minute particles of food go into the anemone. They work together – you protect my back and I'll protect yours. Seeing that working in a biscuit tin in a bar was the first venture into marine science."

Kevin's family has had a long association with the sea. "My father worked in boat-building. My grandfathers, on both sides, fished. So it's going back generations." It's not a surprise, therefore, that he ended up working on an eighty-foot trawler. "We were fishing for any white fish we could get – trawl by day and land the catch in the evening. Now they fish for seven days constant, up to fourteen days, but at that time there were adequate stocks."

If any Dingle fishermen came across an unusual find in their nets, it was off with them to Michael Long's amateur marine biological centre. The pub wasn't difficult to find. "It's the first pub you come to at the top of the pier. When the fishermen would come in of an evening all wet and soggy – they didn't have oilskins in them days – that was the pub they would go into. It's a different kettle of fish now; most of the pubs are for gourmet foods for the tourists. They are not for fishermen to be slugging pints and smelling of fish."

However, the proprietor wasn't doing it just for fun. He kept a notebook which recorded all of the finds brought in by local people. Kevin maintains that over eighty per cent of the rare fish currently in the Natural History Museum in Dublin came from the Dingle area because of two things: the Dingle and Iveragh peninsulas act as a funnel driving the fish from the Atlantic Ocean into the bay; and, most importantly, if anything of interest was located, Michael Long or Kevin would find out and send the fish up to Dublin.

The trips to Michael Long's bar triggered an interest in the natural world which has stayed with Kevin ever since. It started with observing things while out on the trawler, but extended into all aspects of marine life. He admired Michael Long because his interest grew out of everyday things and he had an inquiring mind rather than an academic goal. He explains: "He was one of the old traditionalists and there are few of them left – the people who walked the various roadsides and looked at the herbs and various trees and the birds and knew the whole lot. 'The Last of the Mohicans', we call them. Now, everyone has a degree in environmental science or marine biology or something. I come from that 'Mohican' tradition where you just studied it yourself."

Today Kevin Flannery is one of the country's leading experts on what are described as "invasive species" – non-native marine life found off our coasts. As with Michael Long before him, fishermen along the south-west coast, and far beyond, bring new and strange finds to Kevin at the Aquarium. What's notable about these discoveries is that they are suggesting that our coastal waters are warming. "In the last twenty years, there have been basically lots of what we call Lusitanian fish – fish from Spain and Portugal and Africa. We don't get fish from the Arctic, like wolf-fish. We get mostly warm-water species . . . and these have been turning up more and more off the south-west and west coasts of Ireland."

By the time the fish get to Kevin, they are usually dead. He explains: "The fishermen trawl across the bottom of the sea floor and the fish are all stuffed in together in the nets. Coming up quickly from the depths, the decompression means that they, like divers, get 'the bends' and they don't survive. But when the fishermen are grading their catch, they find these weird and wonderful creatures and they keep them for me. I then have to identify where they are from and what they are doing here. More and more, over the past twenty years, I'm finding quite a lot of species that are warm-water and tropical.

"The most common of them is the trigger fish. A colleague and I have found there is a direct correlation between the increase in sea temperatures and the number of this specific species turning up in Ireland. They can't survive in sea temperatures of twelve degrees or less, but they are beginning to stay around longer. We are finding pregnant females. They are, as we say, beginning to make themselves at home off the west coast of Ireland."

The trigger fish is brightly coloured and about the size of a good dinner plate, around twenty-four inches in length. Thin and long, it usually eats coral and shellfish with the help of sharp prominent teeth. "They have arrived off the Irish coast and have found a feast. We have plenty of crab and lobster and they can quite easily kill these – an *à la carte* menu of shellfish. Their jaws are fused and they just munch the shellfish one leg at a time – disable them and then clean them out. Most of the trigger fish we have found alive have been in the pots off the Maharee islands, off the Kerry coast. The fishermen have to use wide open pots for the spider crab. One of the fishermen found trigger fish feasting on them in every single pot he hauled."

Kevin points out some of the trigger fish which are now in the Dingle Aquarium. In the biggest tank, along with some small sharks, a trigger fish looks as if it is reclining against a rock. What distinguishes it from all the rest is that, when Kevin waves a mobile phone with its screen illuminated, the other fish back off but the trigger fish comes right up to the glass to examine what is going on. It's a fish which does not scare easily.

But because of its teeth, Kevin says the public would be wise to be very cautious. "There's no doubt they are dangerous. Our divers here have been injured by them. These guys will bite and they will take the finger off you. They are quite vicious." He then quips: "They are also quite edible, so if anyone wants to catch and eat them, well, the fishermen and divers would be only too delighted." It is troublesome, though – the aquarium has had to reduce the number of trigger fish in its main tank as they were ganging up on, attacking and killing the other fish.

The toothy trigger fish is just one example of how our seas are changing. In fish markets now it's not uncommon to find silver bream. Red mullet would be another fish which, up to recently, Irish people would have eaten only on holiday in Spain or Portugal. "For every single fish auction in the country, boats are arriving in with them every week – they are quite native now. One time you would only get them when you went down to Fuengirola, or somewhere like that, but now they are quite common. More and more of these species are beginning to turn up."

The big question I have for Kevin is why these types of species are being picked up by our trawlers with such regularity – in other words, are the trigger fish and their friends hard evidence that climate change is taking place in Ireland? As with all previous questions, Kevin is direct. "It's obvious there is a change, or rise, in the sea temperature." And it appears many species of fish are being affected in different ways. "You have a movement north of what we call the gadid species, the cods and the haddocks, because they can't survive in sea temperatures greater than seventeen degrees. And we are getting temperatures of higher than seventeen degrees off our coast." The migration of these traditional fish means of course that there is a food chain available to the invasive fish coming from the south. The habitat is ideal.

All of this interest in fish movement and fish stocks has resulted in a big boost for the Dingle Aquarium. The idea for the development came a few years after Michael Long died and Kevin had taken over his mantle. "I said, rather than sending the fish up to 'The Dead Zoo', or Natural History Museum, for posterity and academics, why can't we get the people of Ireland just to see the fish in their natural habitat?" It wasn't easy – taking five years of hard slog to get the project completed. Why? "At that time in the late eighties, everybody was emigrating from here to Boston. You had two football teams from this area playing in Boston. So it was quite difficult to get the funding. The governments weren't willing."

The first allocation eventually came from Údarás na Gaeltachta – the first time the body was allowed to invest in a tourism project. But there were plenty more turns, including securing permission from the Department of Education to knock down the VEC school standing on the site, and follow-up trips to the Departments of Finance and Marine. Kevin describes it as a "bureaucratic nightmare". It eventually opened in 1996 as Dingle Oceanworld Aquarium, with the considerable assistance of Údarás and local businesspeople. There have been further expansions and today it's an impressive place with an Amazonian display featuring deadly piranha; a shark tank with sting-ray; a touch pool where children can hold starfish; and also a deep-water native tank which has conger eels, large wreck fish and pollack.

Kevin, however, isn't spending much time at the aquarium. His interest now is in tracking the changes in fish stocks. When he thinks back to conversations with trawler men years ago, they were talking of frost and snow on the Blasket Islands. What's also clearly visible is the change in fish movements, which he describes as dramatic. "There is a definite pattern of change in what the fishermen are picking up and not picking up. It can't be all down to fishing inshore. The reason for the stock collapse could very well be put down to climate change. It's become economically unviable for a lot of fishermen now because a lot of these stocks are disappearing."

In saying that, Kevin Flannery is not a prophet of doom and takes quite a jaundiced view of scare tactics which he says are being employed by some environmentalists. He says we have all been there before. "I could become very sceptical – it all started with the acid rain; then we moved on to the hole in the ozone; then we were all going to die in 2000 because there was going to be a computer collapse and our televisions would go off; and now Al Gore has come along and said that everything is going to melt and seas are going to rise and we're all going to die. I think you have to stand back and, you know, everyone has got their interest and some have a big vested interest in creating these things for God knows what reason."

While charting what's happening with climate change though, his immediate concern is for what he calls "industrial fishing" and the impact this is having on the food chain. "If you go into any supermarket, you will see cold-water prawns. Every single one of the supermarkets has these cold-water prawns from up around Greenland. These are the food of the migratory salmon from northern Europe. They are stopping fishermen and putting quotas on them, but they are still not applying the basic principle of stopping the food chain from being affected." His ire is also directed at Denmark for taking "billions of tons" of fish for its pig industry and Norway for taking "billions of tons" of blue whiting for its salmon farms. "If you take a thing out of the food chain, you will have a collapse of other chains, be they cod, herring, mackerel or hake. Being true to the cause, that's what the EU should do – shut down industrial fishing and allow the food chains back. Let's see what things would be like after five years. You can't kill all the chickens and expect to have hens."

When he talks about fish stocks, there's an urgency in his voice. He believes extreme damage is being done right now and this takes precedence, for the moment, over the impending impact that global warming will have. "Climatic changes have occurred through millions of years and will occur again. Animals are quite capable of adapting to these changes and we'll have movements of stocks north, south, east and west. It has given a kick-in-the-arse to people engaged in backyard-burning and the dumping of stuff into the atmosphere and the seas. If it does that, great. At the same time, we have to come down from the high heavens."

While the industrialisation of the fishing industry is clearly taking place, Ireland has also had a hand in fishing beyond what's good for the stocks. I put it to Kevin that we also have to take some of the blame. "Human nature being what it is, any person in an industry, or with a mortgage, will attempt to get the most possible money they can out of it. And that's the way the Celtic Tiger is at this point of time. They are not worried about the ecology, or not worried about anything, until the brakes are put on them by the government or an agency of some sort. If there's a market demand, they will keep doing that. What's happening now is that there's more of a realisation, European-wide, that there are certain things we have to put in place and have to stop. Maybe one advantage of the Celtic Tiger is that the government has become more independent of industry and can stand up."

One thing which is of increasing benefit to Kevin is the improvement in technology and information-sharing. "Initially, when we were starting out, the fishermen were bringing us in these weird and wonderful things. I used to check books in the library and then make reference back down through the years through the *Irish Naturalists' Journal* or the Royal Irish Academy. But now with the internet, access to various organisations and access to data buoys off the west coast, the data has brought us more up-to-date and keyed in to what's happening."

The fact that huge volumes of information have only become widely available over the past number of years has made Kevin cautious about buying in fully to climate change predictions. He wants to see more hard data. "I think there's an over-exaggeration. There are various different things at play here." But he also is clear that his own observations are showing that things are changing, and changing swiftly. "In the last year, I've got puffer fish which are non-native and shouldn't be here. Prior to

2007, we had only ever recorded two or three. In one year alone, I've recorded three of them. Again, they wouldn't survive unless the water was warming."

Which brings us back to those great white sharks. Tabloid newspapers have been only too delighted to report on possible sightings. Yet, according to Kevin, sightings usually don't have much value. "It's very difficult to do sightings in the water, unless you have people from South Africa or Australia who are dealing with them constantly." He's also a firm believer that the fish, immortalised by Stephen Spielberg, will not be appearing off the coast of Ireland anytime soon. "My own opinion is, I don't think so. There is a very intensive shark fishery off Portugal and Spain. If you go into the markets, you still see the fins, you still see the sharks and they are still being targeted. I don't think they'd survive the hooks and lines north of Portugal and Spain to get this far."

Great, I think: there's no chance of the dramatic events of *Jaws* being replayed on Irish beaches. But then comes that unnerving qualification again. "I could be wrong, though, as they are quite capable of surviving here. If they do, it'll only be one or two individuals."

That's reassuring.

4

Storm in a Tea Cup

Problem solving isn't easy. For a start, fixing one problem can often merely trigger another. And then there's the added complication that to solve a problem successfully, you usually have to bring people with you rather than impose a solution on them. Just ask the environmental team at Unilever Ireland, who managed to precipitate a mini-revolt at their headquarters in Dublin. A conflict over cups may sound bizarre, but it's an illustrative story as to how complicated managing environmental change can be. It's also an example of how we, as a nation, are reacting to climate change.

Ireland has an enormous problem with waste. The latest data from our Environmental Protection Agency shows the amount of rubbish going to landfill is rising significantly, rather than being reduced. It's a stark statistic, given that tens of millions of euro have been invested in recycling. If this trend is to change, then everyone is going to have to take action.

Unilever is a colossal global company which controls numerous everyday products such as Lyons tea, Lux soap, Lynx deodorant, Knorr soups, Persil washing powder, Sunsilk shampoo, Colman's mustard, Cif cleaner and not forgetting beefy Bovril. The list goes on, but the firm's website puts it into perspective when it says: "150 million times a day, someone somewhere chooses a Unilever product." The Irish division does manufacture some products like HB ice cream, Quix washing-up liquid and Lipton Tea, but the 200 staff at the Dublin headquarters focus mainly on local selling and marketing.

Unilever says it's serious about reducing its impact on the environment. In September 2007, the parent company trumpeted that it had been recognised as "Best in Class" in its approach to climate change by the New York-based Carbon Disclosure Fund, a coalition of over 300 global investors with $41 trillion in assets. It sounds impressive and the company clearly takes pride in the title, but I think it's fair to say that the public's first reaction to such big claims is suspicion – a sneaking feeling, if not even a tint of cynicism, that there's probably a gap somewhere between word and deed.

The projection of a green image has trickled down to its Irish operation. When I walk into the smart Citywest headquarters, I find an environmental policy statement mounted on the wall which states: "Our aim is to make life better every day for our trade customers, consumers, people and community in an environmentally sound and sustainable manner through continual improvement in environmental performance sourcing, manufacture and distribution." Nice words, but achieving such a goal, as I will find out, is complicated. This is where the cups come in.

In common with many offices around the country, employees at Unilever were provided with the classic white polystyrene cups for tea and coffee. Yet with corporate responsibility and environmental awareness blowing through the market, the white foam days were numbered. The trigger was the decision to establish a voluntary environmental team from the staff, which was charged with informing the board about how their green credentials could be improved.

Eight men and women eventually volunteered for this "Green Team". I drop in to meet them for one of their monthly meetings. The effective leader of the group is Cathriona Murphy, who has worked as an environment manager with Unilever for five years, moving to the Citywest headquarters in February 2007. Also attending that day is Annette Hogan, a management accountant, who only joined the company in June 2007. Tracey Hyde, an accounts assistant, and Deirdre O'Brien, the communications manager, have both been working with Unilever for more than two years. I am assured that there are male representatives on the committee, but unfortunately they are too busy to attend that day.

The group has instituted significant changes; however, a small matter of cups conspired to trip them up. At first, though, they thought they were onto a winner. Cathriona Murphy tells me: "We used to have polystyrene mugs in the business and polystyrene plates for salads and things like that. People don't even know why, but when they see polystyrene, they know it's bad." When the team looked into the facts, Cathriona says their worst fears were confirmed. "It takes a hundred years in a landfill for a polystyrene cup to bio-degrade. So we said, 'Okay, we are going to go down the route of reusable cups.'"

When the wider staff was informed at a meeting about polystyrene's ability to outlive everyone in the company, there were gasps. It was agreed to get rid of the cups fairly quickly. And this is where the real problems started.

First they tried paper cups. Clearly, these were not going to see the twenty-second century. But the Unilever environmental team recognised that paper cups were not going to be the silver bullet either. For a start, you can't recycle them. Even if you wash out the coffee or tea, the adhesive holding the cup together means it has to go to a dump. An average of 4,500 cups a week were heading into a hole in the ground from this office alone.

The next big idea was to examine the possibility of using ceramic cups. Good idea? Not according to Cathriona: "We couldn't get ceramic mugs because, for health and safety regulations, we have to have lids and there was no such thing as a ceramic mug with a lid. We had an incident in 2005 where somebody got burned and that's when lids became compulsory."

With that option ruled out, efforts switched to sourcing a cup which could be recycled. It turned out there were biodegradable options, but this time the price was the problem. While electricity generated by wind turbines is currently the least expensive source on the energy market, being green isn't always cheap. It cost €12,000 a year in waste charges to transfer nearly a quarter of a million cups into a landfill. However, to buy a year's supply of biodegradable cups costs more than twice that. A cool €27,000.

The cost was simply prohibitive. A breakthrough seemed to come when, after lots of digging, they found a company that could provide lids for ceramic cups. The idea was that each staff member would get a cup to keep and maintain. The small problem they overlooked was that not everyone wants to have a personal cup. Why? Because it appears not everyone wants to clean their cup at work. I was intrigued. How could this measure be such a difficulty for some staff? No-one would ever consider using polystyrene or paper mugs at home. If they washed a cup at home, why not do it at work?

In Annette Hogan's view, "The fact that they have cups in their homes, and probably had a cup of tea in the morning and rinsed it out and left it on the draining board, does not translate into the workplace." She opines: "If they had to pay for their paper cups they wouldn't dream of spending that money. Maybe a fiscal measure might work." When I suggest this could amount to a cup tax, the team laugh hard.

I ask the green team if there had been much resistance. Annette's quick-fire response: "Plenty." Cathriona chips in: "Recently we had a tree-planting initiative – giving a tree to each staff member; 144 people were interested. As they came to take their tree, I asked: 'Out of interest, if we were to move away from the paper mugs to ceramics, would you mind washing your own mug?' There was still a bit of a divide, but I was shocked at the number of people who said: 'Ah, that would be hassle.'" A war over cups had begun.

The tussle at Unilever is a struggle which is being replicated right across the country. Companies recognise that "being green" is a quality which consumers are becoming increasingly appreciative of. It's also something which employees are demanding. A recent poll by Ceridian, a global provider of outsourced payroll and HR systems found that sixty-nine per cent of employees thought it was important for their employer to be environmentally responsible and fifty-seven per cent wanted their employer to do more. Some companies are blazing a trail – the internet search engine company, Google, has thirty-two shuttle buses to get its 1,200 employees to and from its office in Silicon Valley, California.

Unilever's environmental team was formed in the middle of 2007. A staff meeting was the first time employees heard about the voluntary body. This was followed up by an e-mail from Deirdre O'Brien. Tracey Hyde joined up after reading it. "It doesn't take any extra out of my day. At home, I've got my own little compost and recycling area. There are things I can bring from home into the office and vice versa."

That concept of applying your personal lifestyle to the workplace also resulted in Annette Hogan joining up in July. She jokes: "I'm a tree-hugger at home I guess. I have my waste segregation. I have my recycling. I do as much as I can at home. It annoys me personally when I see colleagues putting stuff in the wrong bin, especially when the girls and guys have put in so much effort. When you go into the canteen there are five bins and when I see people walking in there and throwing the banana peels into general waste it just annoys me. That's passion, I suppose!"

The team has had successes, paper being a good example. Cathriona explains: "We found we were using 9.3 tonnes of paper per year which works out at around 228 trees. We were using, up until June of last year, one hundred per cent virgin paper. From an environmental point of view, we said: 'Can we not use recycled paper?'" There was, however, a possible price to be paid. A box of virgin paper containing 2,500 sheets cost just under €60, but the recycled equivalent cost just under €72. In the event, the total cost went down because new office systems were introduced – such as photocopying on both sides of a page and discouraging unnecessary printing. Just to keep the pressure on, the team monitored uncollected printed pages and then highlighted the data on a special notice board. The overall effect was to reduce paper usage by twenty per cent, from 2.5 million sheets in 2006 to two million sheets in 2007.

Tackling paper is one of the first steps most companies take – the so-called "low-lying fruit" which, if plucked, can make a big difference in reducing your carbon footprint. But nothing is simple. No sooner have you switched from virgin to recycled paper, then you are being asked about whether bleach was used in the manufacturing process. Without rigorous examination, you can find yourself being accused of doing more harm than good.

The Unilever team then moved onto food waste. While the office canteen had always tried to ensure leftover food was composted, a lot of food waste on each floor was ending up being sent to landfill. A simple plan of putting composting bins on each floor resulted in an additional two tonnes of material being composted instead of dumped.

Energy efficiency was also examined, particularly after a new member of staff suggested Unilever Ireland follow the example of his previous company, which used light-sensors in its offices. Every time you walked down a corridor at night, the lights would click on ahead of you and, most importantly, click off when you passed. Cutting continuous lighting meant a drop in energy use and a decrease in costs. The Unilever team decided to deploy them in a restricted number of areas, like the toilets, where savings were as high as fifty per cent. In a more substantial step, the company also switched to the renewable energy firm, Airtricity.

Yet Deirdre O'Brien admits that not every idea worked out. Take bicycles, for instance. In an attempt to reduce transport emissions, and to improve the health of the staff, the company looked into introducing a scheme whereby staff members would receive €100 if they bought a bike to get to and from work. But it proved impossible to introduce because, as Deirdre tells me, "One, there aren't many cycle paths and, two, we'd put the health and safety of our employees at risk. To cycle to here, where I come from, I would have to cycle through the Red Cow interchange and then down the N7. It's just too dangerous." Car pooling was also looked at but, with staff finishing at different times, it didn't seem practicable.

The question of air travel has also proven thorny. The first step was to work out how much carbon they were generating through business flights. It was established that the company's air travel had produced 180 tons of CO_2 during 2006. It's not a huge amount, given that the Environment Editor of *The Irish Times*, Frank McDonald, "confessed" to having been responsible for twenty-three tons of carbon in 2007 alone.

The obvious place to start was to cut out unnecessary flights. According to Cathriona: "One of the ways we have done this was through the online training courses. A lot of teleconferencing facilities were also put in place."

I am curious to know how cutting down on flights would go down with the staff. It's a truism to say that a direct meeting, in which people interact not just at the conference table but also over the informalities of lunch, can be very productive. Unilever took more of a pragmatic than a dogmatic approach. Cathriona tells me: "There are flights that you can't avoid. You do have to go out and meet people. But there's a lot you wouldn't really need to be there for, like training courses." She lives by her word and on three occasions opted not to travel to the UK for training but undertook online courses instead.

A similar compromise had to be made when it came down to offsetting – the process whereby the amount of carbon generated by an activity, such as air travel, is balanced by an investment in an environmental project, such as growing trees. While reducing the number of flights was the priority, it was very appealing to be able to offset the remainder. However, there turned out to be a big gap between the theory and the reality.

Deirdre continues: "We wanted to reduce the carbon footprint and offsetting looked like the obvious choice. On further investigation it didn't turn out to be so robust." One of the few Unilever products manufactured in Ireland is HB Ice-Cream, a partnership with Killeshandra Dairies in County Cavan – it became the guinea-pig.

Cathriona picks up the story. "We contacted the Green Party first to find out who they were using to offset their emissions. They put us in touch with a crowd in London. We contacted them to see where we would even start. They said: 'Well, *you* can determine where you start and finish, and we'll certify you for that period.' It's strange in the sense that you could say, 'I'll certify for this part of the production and nothing else', when ice-cream by its very nature needs to be frozen, so you should really be considering the amount of energy that goes into keeping it frozen right until it gets consumed. Yet that would be a very difficult thing to do."

The global headquarters is also examining the question of offsetting but, as the concept has yet to be standardised and made easy to use, the option continues to be under review. The board of Unilever Ireland was also put off by the fact that most offsetting schemes appeared to offer proposals such as planting trees in places like Brazil. Verifying what was going on would be difficult.

In the end, after five months of legwork by the committee, the company decided it was preferable to do something locally, even if it wasn't a verifiable offset by the relevant bodies. Cathriona explains: "We got in contact with the Woodland Trust in Leitrim and they are one of the only organisations in Ireland who own their own land, which ensures that the trees that are planted are going to be in the ground long after we are all gone. We planted a tree for every ton of carbon. It wasn't really an offset because we didn't really go down that road, but it was something good."

Because their green dreams have not always been realised, the environment team has become quite strategic. When recycled paper was first introduced in 2005, it caused all sorts of problems, including jamming up photocopiers. Within a short space of time, it all had to be replaced with virgin paper again. When improved recycled paper was sourced the following year, it was introduced on a phased basis without informing all of the staff. The fear was that colleagues, after a bad experience the first time, would actually be looking for problems. When no problems arose, it was easier to then inform all of the staff about the changeover which had already taken place.

In Cathriona's view, what has also made it easier is that her colleagues are becoming more environmentally aware, particularly in the home. "You have your green bin in your house. Now you're getting brown bins as well for composting. The government has done a lot with the Power of One, which is really hitting home. It's just unfortunate that this mug issue has become such a massive thing for us."

Just before completing this book, I contact Deirdre O'Brien to see if the war of the cups has been resolved. As it turns out, the ceramic cups were introduced in February 2008 and the initiative has been a resounding success. "Most people are putting their mugs in the dishwasher or rinsing them out. So, despite initial grumblings, people have welcomed the new ceramic cups and the feedback, so far, has been positive." The next challenge, no doubt, will be to assess how eco-friendly the dishwasher is.

Even when the cup war was rumbling, the environmental team had made significant progress and it was appreciated by their colleagues. Internal polls bear that out. When employees were recently asked whether they agreed with the statement, "I am proud to work for Unilever Ireland", eighty-three per cent agreed, up from seventy-six per cent the previous poll.

There's also a competitive edge to all of this green activity. Many companies now issue news releases on how green their business is. Up and down the country, there are big companies and small ones trying to introduce more eco-friendly practices.

Take Diageo, for example, the parent company of Guinness. By 2007, it already had a zero landfill waste record for three years. This was partly possible because some of the liquid waste from beer production is used as a fertiliser to help grow willow trees. Like Unilever, it is now offsetting air travel and has a renewable energy-only contract for its breweries. It was also only the second company in the state to secure the latest energy management standard, IS 393.

In September 2007, Tesco Ireland opened a new shop at Celbridge in County Kildare with key innovations such as ensuring that its lights were sensitive to daylight and dimmed on bright days. It also had special fridges which ensured that cold air didn't spill wastefully into the aisles. And it has ambitious targets for the future, such as becoming the first Irish retailer with one hundred per cent recycling by 2010. It's spending €30 million on cutting energy consumption by fifty per cent in its ninety-five shops here. It hopes to have the most environmentally friendly store in Ireland by 2009, built solely from recyclable materials.

Another company leading the charge is Musgraves, the wholesaler which provides food to independent retailers. It claims to be the first Irish company to obtain all of its electricity requirements from a green source. By 2006, it already had a waste recycling performance above sixty-five per cent. Its award-winning Cork headquarters has an energy management system which means usage is a quarter of traditional structures. Given its heavy reliance on delivering foods by road, it is focused on eliminating five million kilometres of lorry movements from Irish roads each year by upgrading its distribution network. The work was recognised when Chambers Ireland awarded Musgraves the best-in-the-class award for corporate social responsibility in 2007.

Internationally, Unilever says it's ready for the wider challenges of global warming. Its sustainable development report for 2006 is chock-full of plans and initiatives. It has formed a Greenhouse Gases Working Group of senior managers; CO_2 emissions were cut by 4.2 per cent; 14.8 per cent of energy came from renewable sources. On the product front its detergents can now be used in washing machines at temperatures of thirty degrees; a re-design of its shampoo bottles in the US led to a saving in plastic resin equivalent to a reduction of fifteen million bottles.

But the company has not always been praised for its corporate practices. Type "Unilever" into the website of Greenpeace International and up pop three pages of less-than-positive articles. There's condemnation of Unilever's links with the palm-oil industry and ongoing deforestation; and criticism resulting from its perceived promotion of genetically modified crops. There is, however, also grudging praise for Unilever's decision to ditch its existing refrigeration systems in 2004. The company's response to these issues is that, on palm-oil, it has formally committed itself to sustainable production practices while, on GM crops, all of their food products are GM-free.

And everything is under review. Deirdre O'Brien tells me that the buzzword in the wider corporation is "brand imprint". "Brand imprint helps brands measure and identify the social, economic and environmental impact they cause. Let's take Persil – or Omo as it's known internationally: it's a billion euro brand. We've looked at it in a totally new way – with 'green goggles' on – to see how we can affect the packaging, the carbon footprint, the entire thing. And all brands are slowly going through this process. Persil is changing. It is now greener."

But if climate change is the greatest challenge for our generation, because it puts future generations at significant risk, what will happen when Unilever Ireland completes the easier stuff – i.e. reduces waste, cuts back on flights and lowers energy use? While the staff have been out doing the hard stuff to date, much tougher choices will ultimately have to be taken by management. For example, if the government strategically decides not to invest in public transport in their area, will the company move? Will they employ only those staff who are prepared to abide by strict edicts on environmental practice? And what about the Unilever corporation? Will it be prepared to discontinue popular brands because their carbon footprint is simply too big? Will it axe jobs in one country because it can be done more environmentally elsewhere? Ultimately, is it prepared to lose money for a period of time to be green?

What's clear from Unilever Ireland is the staff's dedication and interest in making a difference. It's also evident from their experience that it's a complicated and, at times, frustrating job. As things move on, these challenges are going to get bigger, not just for this company, but for all companies. If the science is right on the speed and impact associated with climate change, extremely radical, difficult and expensive decisions are going to have to be taken. And it won't be easy – judging from Unilever's storm in a tea cup.

5

Irrational Behaviour

Kevin Myers raises a teaspoon from its saucer and, with something of a flourish, places it on his kitchen table. It is a bright January afternoon and the often-controversial columnist is explaining his inherent scepticism about the concept that mankind is responsible for global warming. "That teaspoon has just been moved from its saucer, because I've moved it. We do not have that level of certitude, or knowledge, about what's causing this warming to the planet now."

It's no surprise that he isn't overly impressed by the overwhelming scientific view that emissions from cars, factories and intensive agriculture are responsible for climate change. His first line of attack is to point out that so-called experts have been wrong before. "If you had gathered all the great thinkers of the world together in 1400 and asked them a simple question: 'Does a heavy object fall faster to the ground than a light object?' They would have said, without question, that a heavy object does. It's not true – it's gravity."

If they were wrong then, he opines, they could be wrong now. "Science discovers itself every Thursday. There are new revelations about the origin of the world or the newest smallest particle. Science undoes itself the whole time. It seems to be irrational behaviour for mankind to change all our behaviour so totally, because of a scientific theory which might not turn out to be correct." Rational thought and logic are recurring themes throughout our conversation.

In his regular *Irish Independent* column, Kevin previously went beyond simply doubting the veracity of the link between climate change and human emissions. In one article he stated: "There is nothing mankind can do about global warming. Nothing. There is not even any proof that mankind causes it, merely evidence that it exists." Those who argue there is a link were dispatched mercilessly. "The eco-priests of our modern lunatic sect, intoning their fluent gibberish, dominate all discussion about the environment." He concluded: ". . . the real source of global warming [is] the sun, all 2,000 trillion trillion tons of it."

However, by the time we sit in his kitchen in Ballymore Eustace, with a wonderful view of the Wicklow mountains, he has taken a step back from blaming solar activity. He describes himself only as a "sceptic" on the question of humanity being responsible.

What irks him most, it seems, is his perception that the public is being forced to conform, when an element of doubt remains. "In many regards, the obsession with the environment is rather like a new religion – you conform, or you are some kind of heretic and unworthy. I'm a sceptic by nature. And I'm just sceptical when a new conformism arrives, particularly when it's surrounded with a shrill element of hysteria, and an intellectual imprecision, which is more to do with emotion than scientific fact. I'm not an expert on this, as you know, but I get nervous when I see mobs – when crowds gather around a central unifying belief."

His central thesis is that the public is signing up to this concept, even though it's unproven, and they don't fully understand it, because human beings have a primal need to belong to a group. "We are a very flawed species. It's why religion is so popular around the world – we have a requirement for a greater unifying belief. We are primates. We're not rational people. We're not Dr Spocks. We have an emotional need, a psychological need, to cluster. The cluster point at the moment is the environment."

This takes him back to the evidence for believing we are responsible for climate change. "The environment very well might be in danger. All the evidence suggests that terrible things are happening all around the world. Yet there are a minority of people who say 'this could just be solar activity'. We don't know. There's been no proven connection between the carbon dioxide and what's happening to the world. The history of the planet is so complex as to be beyond any individual's understanding."

If Kevin Myers is nervous about the mobs, he's downright hostile to the leadership of the eco-movement. First in the firing line is the Green Party. "I don't like the Greens at all. I see so much of the old Catholic Church in the Greens. There's a kind of wild dogmatism which makes me nervous. It does not mean you are virtuous, just because you believe in the environment. It does not raise you above mob instincts. I see the potential in the Greens for all sorts of nasty totalitarian possibilities."

It's classic Kevin Myers. Over many years, he has devised a canny ability to construct highly controversial arguments on topical subjects which result in the Letters to the Editor pages filling up. Yet it's also a truism to say that Kevin wouldn't be well paid for so many years if he wasn't delivering readers to *The Irish Times* and, now, the *Irish Independent*. It seems that while his readers don't necessarily agree with what he has to say, they do enjoy the caustic, sometimes shocking, way he phrases things. Clearly, we like people getting it in the neck.

While we are talking, he usually avoids making eye contact, looking out the back kitchen window while warming to his theme. The green movement is a major concern. In the main, it apparently stems from how they communicate their message. "There's an appearance of a blithe smugness; that they are morally superior to the rest of us. You will get such blithe smugness in insipient moments of any universalist totalitarian movement – 'we know the answer and we're morally better'. I see in the green movement the possibility of totalitarianism. These are very early days. It does not necessarily mean it's going to go down that way. But when you have a movement that is so total in its belief; that is so certain that it's right; that is not grounded in the preciousness of human life but in the preciousness of the planet or animals; well, then you are looking at something which is potentially dangerous."

Kevin believes in equality; he's happy to lash any organisation. He tells me governments are also responsible for dealing in fear rather than fact. The emergence of the AIDS crisis in the mid-1980s provides a good example. "Twenty years ago we were all told that we were equally at risk of contracting HIV. This wasn't true. It was a lie. The people who were most at risk in Europe were male homosexuals, who were having anal sex, and intravenous drug users. Yet the health bodies around Europe agreed on this central lie. So it does seem that even intelligent, rational people will subscribe to a fantasy in order to encourage believers."

There is, he maintains, a parallel between the way religion operated in the past and the green movement of today. "You can't logically explain religion because it's got nothing to do with logic. It's to do with fear – fear of death and fear of the consequences of not having lived a good life. And these people are behaving like any people starting off a religion. In a perverse way, it's a logical thing to do – get people on board out of terror."

In recent times, many religions have spoken about the need to act on climate change. Indeed, the Catholic development agency, Trócaire, focused its 2008 Lenten campaign on the issue of how global warming was impacting on the poorest in the developing world. Kevin derides such church involvement as nothing short of a survival technique. "If the Catholic Church thinks it can become more popular by preaching environmentalism, that's what it'll do. It's logical. They want people to supply them with money; they want people to go to their churches; they want to be liked. Personally I think it's lunatic for the Catholic Church to associate itself with political causes. If you are a Christian, that is not why Jesus died on the cross; that is not the purpose of the Last Supper. It might work though. But I feel, God forgive me, a small sneer of contempt settling onto my lips when I see the Catholic Church scrambling, rather desperately, on board something like the environment."

Even when speaking, Kevin Myers has a tendency to be emphatic on his view of what's happening in the world and then go for the jugular vein of any perceived opponent. Yet despite his harsh, sometimes demeaning words, you could suggest that the gulf between his position on climate change and, say, the government's, is not that wide. For a start, he accepts that he's not an expert on global warming. He recognises that the overwhelming scientific opinion is that humans are responsible. He does not dispute that the earth is heating up or that the most cautious predictions suggest that the consequences could be disastrous.

Given that, I ask him whether he would not accept what's called the "precautionary principle" – that it's logical to act now to stem greenhouse gas emissions, rather than do nothing and wait for the probable confirmation that global warming is man-made. His answer? "I would go along with that." But then there's the inevitable qualification: "I see no reason why we should be foolish or reckless, but we have to accept that there are extreme limitations on our ability to change anything. When I accept the precautionary principle, it's because it's not going to do us any harm. It might not make much of a difference, but it makes us feel better, and that's not a bad thing."

That doesn't seem to be the most logical of reasons to take action, so I ask him to explain why, in his view, there are extreme limitations on mankind's ability to change things. The example he chooses is China. "The 550 coal-fired power stations that the Chinese are building suggests that the limitation on a two-litre car in Ireland isn't really relevant. These coal fires cannot be put out. They produce more CO_2 than all the motor cars in North America put together. Intellectually you know that the argument comes to a full-stop there."

China is building around two coal-fired power stations every week, and is believed by many already to be the world's number one polluter. It's also true that Irish emissions are but a fraction of the Chinese total. Yet that's only part of the story. If one looks at the issue from a per-capita basis, the average Chinese emissions are tiny when compared to those of the average US citizen. Embarrassingly for Ireland, our emissions are, on a per-capita basis, the second worst in the EU and the fifth worst in the world.

When I suggest to Kevin that Ireland has to reduce its emissions before talking about China, he changes tack. "I wouldn't say anything to the Chinese government. I would say: 'You are entirely right. Your duty is to advance the welfare of your citizens. It would be insane for you to attack your economic growth in pursuit of a theory that is not yet confirmed.' Even if it were confirmed, it wasn't China that wrecked the world's environment, it was the west. So why should China lie in poverty when the western world enjoys riches? There's no logic there."

Yet the Kyoto Protocol factors in this issue of historic emissions. It's why the thirty-five developed countries, plus the EU, collectively agreed to reduce their emissions between 2008 and 2012, by five per cent below 1990 levels. They committed themselves to this even though no specific targets were placed on China and the other "Big Four" emerging economies – Brazil, Mexico, India and South Africa. Yet it is very contentious. The United States decided not to sign up because the developing countries were, it argued, being let off the hook. Now negotiations on a successor to the Kyoto Protocol are underway and one of the crunch issues is how to ensure that all countries, including the US, agree to cut CO_2 by fair and defined amounts.

For Kevin Myers, however, the per-capita argument leads to a far more depressing outlook. "So what are we doing worrying about what Ireland does when it's not going to make any difference? We *know* it's not going to make any difference. It's a piety. On the one hand, I know pieties are nice things and make us feel better, but this is a meaningless piety." He then projects a rather baleful vision of the future. "I don't think there's very much we can do about this. If the damage has been done on the lines people have said, then it's already too late. We have to recognise this possibility that we have stepped into the funnel and that the planet is doomed. It's possible this is the case."

This grim conclusion is based on what he sees as the total inability of the governments of the world to do anything other than talk platitudes. In Ireland, as in most other places, the amount of CO_2 being generated is far above the level it's supposed to be at under Kyoto. This suggests to Kevin that the politicians are not serious about the commitments, but are serious about getting back into office. "What the political world is talking about is not saving the planet but getting re-elected. You have to take on board the piety, because otherwise you won't get elected."

In his cynical view, the electorate feeds that political outlook. "If you say to the electors of Darndale or Sandycove, 'What's going to happen is this: we have to make serious sacrifices to save the planet; we are going to prohibit the use of motor cars; we are going to allow you two hours of electricity a day; you can only have one light bulb per house' – you will get just one vote from a gibbering lunatic in a garret in your constituency." It's a bleak scenario. "We know the politicians are not serious about the threat the scientists are talking about. If they were, it would have all been resolved in Kyoto. They would have said: 'Okay, we're going to close down western Europe and North America.' But politicians know if they try to do that in a free world, they wouldn't be re-elected. They are not honest in their language, nor are most media commentators."

Kevin's view is that the public need to face up to the fact that there is no such thing as a global community. While there is lip-service, through the UN, to the idea of the collective good, the harsh reality is that it's national self-interest which determines policy. "It makes a great deal of sense for Ireland to import things from China because it means there isn't a steel mill up the road from me. It's a rational decision – I let the Chinese get the toxins and the dirt." If things are to change, it will require a political revolution which has never occurred before. A rebellion in which the voters are told the truth by their politicians, before an election, and then receive a mandate to take those hard decisions. "If a party does that, and is elected, it will be the first poverty party ever elected in the world. If it said, 'We are going to cut your incomes and we're going to increase taxes to save the environment', it would be the first time in history that mankind itself has shown itself to be a virtuous species."

Isolated is a place Kevin Myers often finds himself. Although he's a graduate of University College Dublin, he was born in Leicester, England, and this lineage is something he feels has set him apart in Ireland. He does not view it as a negative thing. Indeed, this position of "outsider" is something he believes often offers him a unique opportunity to evaluate Irish society, because he's not wholly a part of it. He can take on taboo subjects or tackle a perceived liberal-consensus. And he does get stuck in regularly. Among the sensitive topics which he has waded into are feminism, Travellers, immigration, republicanism, Islamic fundamentalism, and the war in Iraq. On occasions he has grossly over-stepped the mark. For example, in 2005 he used the word "bastards" while writing about social welfare claims lodged by "unmarried mothers". He apologised unconditionally shortly afterwards. His articles, at times, may well oscillate between dripping sarcasm and full-scale assault but, when you meet him in person, he's a welcoming host, self-deprecating, irreverent and good company.

When it comes to the subject of climate change, he clearly feels there's plenty of opportunity to, as he puts it, cast doubt. Throughout 2007, he employed parody in his

Irish Independent column to deflate dire predictions on global warming, as in this one: "Céad míle fáilte to Ireland, 2050. Take a look around at your new home. It's a bit of a squeeze – we were always a small country but you might have heard we've lost quite a bit of land over the past few years. Flooding, of course, and then a few nice towns have fallen into the sea, too. Poor old Wexford. You should try and squeeze in a visit to Waterford soon – it's not long for this world. Bring your malaria tablets, though; the mosquitoes in that part of the world are vicious little yokes. We used to call it the sunny south east; now it's just 'New Wexico'."

There was barely concealed delight when he got the chance to have a cut at Nobel Peace Prize Winner, Al Gore. "In his heart, Al Gore knows he speaks sanctimonious, forked-tongue flannel, for recently released figures reveal that his home – twenty rooms, eight bathrooms – consumes more than twenty times the amount of electricity of the average US dwelling. The Gore home's energy consumption has actually increased by fourteen per cent since he made his preachy, hand-wringing, time-bombing, tail-spinning film. Imagine the outcry if President Bush had been caught out in such rank and flagrant hypocrisy." The end of the year offered him the not-to-be-missed window of launching an assault in the *Irish Independent* on delegates attending the UN environment summit in Indonesia. "Rio. Kyoto. Bali. That's environmental conferences for you. They always occur in sunlit places ending in vowels, and with a consonantal component of no more than fifty per cent. They're never in vowel-light locations like Nitvinggen or Bblarrgh or Quivdansk, where summer lasts a few hours some time in June, and where the locals spend their long winters rummaging through their clothing of animal pelts, popping lice with gnarled, nutshell fingernails, and musing vowellessly."

As he must predict, the letters of criticism quickly follow: "Given Mr Myers's tenuous grasp of scientific argument, and his use of ranting emotionalism, I would not expect that PhD to arrive swiftly. Perhaps, the Flat Earth Society could present him with an award for standing up to modern voodoo science!" But it's not all criticism. One reader wrote: "Thank heavens for Kevin Myers. Not for the first time he's proved to be an oasis of sanity in the barren wastes of bankrupt thinking; this time on the environment!" Another said: "How fortunate we are to live in Ireland, where there are people who are capable of seeing through the rubbish climate scientists come out with." The letters illustrate that there are sceptics out there and, in Kevin Myers view, society is better for them.

He clearly knows how to identify trouble and position himself in the middle of it. Maybe it's a skill honed while reporting from various hot-spots. He was in Northern Ireland during the height of the conflict in the early and mid-1970s. Later, he sent despatches from wars in the Middle East and Bosnia.

Sometimes you wonder, though, if the constant invective takes away from a possibly valid point. This crops up when he talks about how humanity has a constant urge to "do something" when confronted by a major problem, such as climate change. His point is that, quite often, such an act can make the problem worse rather than better. "Africa is looking at a catastrophe. And the rational thing for the world to be doing is saying, 'We have to produce water, stop desertification and control population.' What are we actually doing? Well, Bill Gates has announced plans to abolish malaria in Africa. If we are going to have a reduction in child mortality, which I want, you have to accompany it with massive investment. There has to be infrastructure: cities, roads and education. You have to create a civilisation in Africa which will help it absorb the children who are not dying from malaria. Otherwise the elimination of malaria in Africa is a vain project. We tend not to think about the consequences of what we do; instead we project our morality mono-dimensionally on the planet. An awful lot of good deeds are for the gratification of the person who does the good deed. You have to beware of that emotion."

It's certainly a point worthy of debate but Kevin Myers can't resist the opportunity to shock. With a twinkle in his eyes, and a quick verbal flag – "you are going to love this" – he goes on to assert that malaria is one of Europe's greatest friends as, if millions of Africans didn't die of the disease, then they would be crossing our borders because Africa can't offer them anything, and Rome would end up looking like downtown Lagos.

It's clear that while Kevin Myers may well be out of step with the scientific and political near-consensus, his ongoing onslaught on the eco-movement does garner some public support. Maybe it taps into the Irish non-conformist culture – we don't like being told what to do. However, the fact that he continues to insinuate that climate change could still be some form of elaborate hoax or a crack-pot theory enrages many environmentalists. Given the amount of work that needs to be done, and the absolute requirement for public buy-in, such distractions are viewed as not just irresponsible, but dangerous for the planet. His seeming U-turn on the solar issue is scoffed at.

Yet it has to be faced that, on many occasions, Kevin Myers has a valid point. The UN says it is "incontrovertible" that climate change is caused by mankind, but its own advisory committee, the IPCC, can only say it's more than ninety per cent sure. There is still a huge gap between the narrow time-frame for action, as suggested by the scientists, and the absence of an urgent response from the world's governments. At home, Irish governments have consistently pledged themselves to act on climate change, but our emissions are nearly twice where they are supposed to be. People like Bjørn Lomborg – author of *The Skeptical Environmentalist* and, most recently, *Cool It* – suggest that while massive resources are about to be invested in reducing emissions for unknown results, maybe it would be better to deploy such resources on fighting HIV/AIDS or providing clean water to desperate people today.

One question which his critics unite around is: if he's so smart, what positive suggestions does Kevin Myers have? After he has criticised, condemned and excoriated everyone during our conversation, I ask him what solutions he might have to offer. But there are none. "I don't have answers. I'm a natural sceptic – I haven't got a solution. I don't know." It is certainly an open and honest answer. But I suggest it is an untenable position, given the venom and apparent glee with which he sinks his teeth into everyone else who proffers a possible way forward. "It's easy in one sense that I don't have to martial any scientific argument. But I'm not trying to present an argument, I'm just raising doubts. I'm not trying to win any argument. I can be persuaded that I'm wrong." An alternative suggestion might be that he's simply opposing the majority position for the joy of kicking the establishment and getting profile for his column. "I don't think I'm doing that. Am I a professional contrarian? No, but I need further levels of evidence and common sense than those which have been proposed to justify the green agenda in Ireland."

It's a view he holds steadfastly – the right to criticise everyone, sometimes viciously, without offering an alternative. The problem is that, after a while, it makes him uniquely easy to be dismissed out of hand – good points with the bad. To borrow a phrase from a totally different context, as Mandy Rice-Davies put it: "Well, he would, wouldn't he?"

6

A Pacifist at War

Global warming has become mainstream over the past two years. Everyone seems to be comfortable talking about carbon footprints, eco-friendly houses and the plight of the polar bear. For green campaigners like Pat Finnegan, it's a relief that the penny has finally dropped on just how serious the problem is. However, he argues that the information has been out there for a long time – it's just that no-one paid any attention.

When giving public talks, he likes to use a quote from a US academic, published in the high-profile magazine *Fortune*: "There is a very important climatic change going on right now. It is not merely of academic interest. It is something which, if it continues, will affect the whole human population of the world, like a billion people starving. The effects are already showing up in rather drastic ways." After revealing the text, he asks members of the audience to guess the year. Most people suggest the quote is from around the year 2000. No-one ever gets the real date: 1974.

In that decade, there were several warnings, from people like the maverick British scientist Dr James Lovelock, that industrialisation and intensive agriculture, coupled with an expanding global population, were simply unsustainable. Something was going to give. By the following decade, meteorologists and climatologists were becoming increasingly concerned with trends occurring in the earth's atmosphere. In 1992, the world's politicians bought into the concept at the so-called Earth Summit in Rio, Brazil. The problem, it seems, is that the political drive, public awareness and media coverage quickly dissipated. The same thing happened in the late 1990s with the signing of the Kyoto Protocol. Essentially, many people found it to be an interesting subject but something which had little, if any, direct relevance to their daily lives.

It wasn't until the end of 2006 and beginning of 2007 that global warming was on everyone's lips. Three things seemed to happen in quick succession: the publication of a report into the financial implications of global warming by the former chief economist of the World Bank, Sir Nicholas Stern; the release of former US Vice-President Al Gore's film *An Inconvenient Truth*; and the dramatic predictions contained in a series of reports from the UN's Intergovernmental Panel on Climate Change, or IPCC. Things would never be the same again.

Pat Finnegan straddled all of these developments. In the 1970s he was tracking the emerging environmental science while working as an organic farmer. In the 1980s, he became politically active with the Green Party. In the

1990s, he dedicated himself to campaigning on fighting climate change. In 2007, as a reviewer for the IPCC, he was – with only a little bit of exaggeration – a part-winner of the Nobel Peace Prize. The outsider had become mainstream too.

However, the spark for Pat to get involved in the environment came much earlier – at secondary school. "My family moved to England when I was very young so I was brought up in the English education system. I just hated it, until I started doing A-levels in biology at a school near Oxford. There was one brilliant teacher and ecology was his thing. Everything I had been taught up to then seemed to me to be against nature. Suddenly here was something I was being taught which actually took cognisance of nature and the way it really works."

Pat went on to study agriculture and forestry at Oxford University. He turned down zoology – "I didn't want to keep gorillas" – and also decided against botany – "I wasn't sure what botanists really did." So what was it like at one of the greatest universities in the western world? "It was awful." Part of the problem was that most of his fellow students – fourteen in all, of which one was the department's first female – were from the aristocracy and attending college in order to manage their 4,000-acre estates in Hampshire. "The one thing really worth learning was that I did not want to do agriculture and forestry in the way it was being taught there. To me this was completely contrary to having a sustainable planet in my lifetime, never mind the next century."

The word "sustainable" is in common usage today but, in the 1970s, it was a concept very few people were aware of. Yet some signs were emerging. The year Pat graduated, 1972, was when the United Nations decided to hold its first inter-governmental conference on the environment in Sweden. More importantly for Pat, many ordinary people decided that the world's politicians couldn't be trusted with saving the planet and so decided to crash the party. "The green edge of the late sixties' generation just showed up – a bit like Woodstock. People came from the United States but mainly from Europe . . . saying, 'This is my right, according to the UN Declaration of Human Rights. I'm supposed to be involved. Here I am – let me in.' No-one knew this was going to happen. I mean, this was 1972; there were no mobile phones or computers; it was purely organic. But it was a problem for the UN as this had never happened before at a big conference. They had to take a decision – do we kick them out or let them in? Thank God it was Sweden – the police said, 'Let them in.' So 1972 was where the click happened for me."

Armed with a degree which he felt was almost valueless, Pat Finnegan pondered his future. He firmly believed that the planet was heading for big problems if things continued as they were. The emergence of Greenpeace seemed to indicate that others were concerned too. But what action to take was unclear. With no activist guidebook in existence, Pat decided to become a gardener and try to work it out while digging the earth. "I thought: 'I need time to think about this. I'll go off and try to be as self-sufficient as I can, and that'll give me enough time.'"

For nearly a decade, Pat travelled between Greece, Italy, France and Ireland. By 1981, though, he was ready to settle in Dublin. However, it was a very precarious existence. He made a living by hiring himself out as a gardener for a day a week to middle-class families with half an acre or more. "Gardening in 1982 was not a trendy thing. It was not a cool thing. So I didn't get too much money for hiring myself out at the time. I'd only be called in when the gardens were really gone, possibly because the gardener they kept on had died, and it was going backwards. Sometimes I felt like a bit of a vampire."

The deal which would be struck was unusual because Pat's labour was effectively bartered for either accommodation or use of part of the garden. "I would cut a deal whereby I would lay out a garden design plan. I would say: 'This is where I believe we should start. In the meantime, that other area is just going to have to stay the way it is. But if you let me have the use of it to grow my vegetables, you get the weeds out for free.' Bob's your uncle."

To make ends meet, Pat ended up working virtually every hour of daylight there was. When he did get the chance for downtime, he would be reading up on the latest environmental developments in *The Irish Times* and *New Scientist*. However, he didn't feel he had time to become an activist. "I had my nose one foot off the ground. My life depended on how my onions grew and carrots grew." That changed in 1986 with the explosion at the Chernobyl nuclear plant in Ukraine. Sitting outside the political realm didn't seem an option anymore. At, home, the Progressive Democrats were formed with the stated intention of "breaking the mould" of Irish politics. As the 1987 election loomed on the horizon, Pat Finnegan decided to act. He joined the Greens. "Breaking the mould of Irish politics? Okay, I agree the country desperately needs it. But, if we are going to re-jig it . . . not this way, lads! In fact, you're only making it worse. The Greens needed everybody they could get."

The system of gardening he had astutely cultivated for fifteen years collapsed in 1996, with dire consequences. "Someone whipped the rug out from under me . . . I suddenly found myself without anywhere to live at short notice in the middle of winter. With that lifestyle, when you have nowhere to live, you have no job . . . The Council eventually housed me as a homeless person, a pretty rough thing, but it was fine compared to being homeless. I had never been unemployed before in my entire life. It was a big shock. Homeless and unemployed? Can you imagine? My possessions went to the four winds. It's not that I'm possessive, but I am very sentimental. Thanks to Dun Laoghaire-Rathdown Council – and I'll never forget them – I got housed there."

If 1996 was the *annus horribilis*, things began to come together in 1997 in a variety of ways. Firstly, he was accepted onto a master's degree course in equality studies at UCD. While many students begin courses with only a vague idea of what they will focus on, Pat was very clear. He had been following plans by the world's governments to try to negotiate a treaty in Japan which would curb the growth of greenhouse gases. The big questions were which countries would move first, by how much and by what date. "I knew that the Kyoto talks were scheduled for December 1997. I showed up in Belfield with my abstract paper written – how to share out a global carbon budget equally, bearing in mind history and historical responsibility. I had such a fabulous year. There was me at the age of forty-four, doing what I didn't get the chance to do in Oxford – showing up keen."

Pat became immersed in the issue of climate change, picking up distinction grades as he went along. By July 1998, it was time to write up the thesis but, on the way to the library at UCD, he fatefully decided to stop and have a coffee, a cigarette and a read of the paper. "I got to page eleven, and there was this little box on the bottom right-hand corner. It had the Custom House logo on it, saying: 'The Government has received a consultants' report which will form the basis for Ireland's first National Climate Change Strategy. Submissions sought by end of July.'" Pat decided to submit a proposal, as it might help him refine his own thinking for his thesis. It didn't quite work out like that. "In my mind I allocated it a max of a week. In fact, it took me the whole of July. I did nothing else. I've never worked so hard with my brain in my life. But, boy, I'm glad I did it."

On foot of his submission, Pat was invited to a consultative meeting with then Environment Minister Noel Dempsey. Armed with his thirty-two-page document, attitude and trademark shoulder-length hair, Pat Finnegan went to his first encounter with a minister. "The great and the good from Irish industry were all there, senior principals from seven or eight government departments also. After a while, I said, 'I want to speak here – it's a consultation isn't it?' Fair play, I actually was given the chance. That was the first time anyone in the Irish policy arena met this weird hippy Pat Finnegan. I actually got a taste for it. I started to fire off stuff to every consultation."

Pat was also trying to live his life in as sustainable a way as possible. For example, he had never taken a plane flight in his life, preferring to go by rail and ferry. Not having a car, he uses the DART to carry his recyclables to a bring centre every six weeks. It's here where he often encounters a confused public reaction to global warming. "I have to get on the DART in Blackrock with several black plastic bags containing paper, cardboard, a small amount of plastic packaging and the odd tin. It's considered a bit weird, but that's fine. What I have observed, many times, is a bloody great Range Rover showing up, 4.2 litre . . . and the owners throwing five or six green bottles in one skip and a few newspapers in the other. I just suspect there are people making these trips nearly every day. If you look at the planet, any common-sense person must agree that it doesn't make any sense."

This is a constant theme with Pat – the apparent time-lag between the scientific evidence and the political/public response. "For those of us working at the policy coal-face, what we all find terrifying is the delay between what the science is clearly indicating and how the politicians and public react." The more he knows, the more concerned he becomes. Yet this is not replicated among the public which, it seems, still finds it hard to get to grips with climate change impacts which are decades away. Pat finds this difficult to understand. "Most young people have signed up to a twenty- to twenty-five-year mortgage. Why tell me that understanding what things might be like in twenty-five years is beyond you? I believe people predicate decisions on 'business as usual'. The key message from climate change is that business will never be as usual again. Climate change does what it says on the tin – everything will change."

One point of optimism for him is the growing international pressure being placed on governments by activists. He got his first taste of this coalition when, in 2000, he travelled to a UN meeting on climate change in the Hague. His trip to the Dutch city, under the umbrella of Friends of the Earth, involved a forty-eight-hour trip by boat and train. The fact that he was prepared to take such an arduous journey, instead of a two-hour flight, ensured that his very first press release got him into studio with Pat Kenny on RTÉ Radio. As he ruefully comments, it was the first and last time he's been there. "Why the Hague was a turning point for me was that I met people from Climate Action Network, the global activist forum. It was the first time I was outside Ireland in twenty years and, when I got to my first CAN strategy meeting . . . oh, it felt like I'd come home."

One reason he quickly became enamoured with CAN was that he got to see its power at first hand. He watched as a US proposal, which activists didn't like, was seemingly going to be supported by the EU. Overnight, CAN worked on an article and published it in its newspaper, *Eco*, the next morning. Pat maintains that this changed everything. "It contained a complete re-interpretation of the US figures. By the time the EU held its lunchtime coordination meeting, they decided to oppose the US suggestion." A short time after the Hague meeting, Pat fell out with Friends of the Earth and set up his own organisation – Greenhouse Ireland Action Network, or Grian. It's the only Irish activist group aligned with CAN, an entity he feels has a huge influence. "Governments have a high turnover of personnel, but we don't. We're therefore becoming more expert and that's where the respect and relationships happen with officials."

One question about his work which does make him bristle is the appropriateness of taking plane journeys to participate in UN negotiations. In December 2007, the UN met in Bali, Indonesia, and some cynics suggested it was more about winter sun and shopping than saving the planet. "I never travelled anywhere until I started trying to fix this problem. I travelled overland to my first three UN meetings: the Hague, Bonn and Marrakesh. It's a work thing. If I wanted to risk getting negative, I would just ask the critics how many times they've flown for fun? It's also a global problem. For instance, the next session is going to be in Thailand. I was talking to this Custom House guy and I said: 'Could it not be in Bonn?' He responded, 'Then the Thai delegation has to fly to Bonn. What are you talking about?'"

If the issue of air travel elicits a passionate response, a question on Al Gore stumps him. Given that there's been some controversy over the man and his film, I ask Pat directly: what does he think? For twenty seconds he ponders his response. When he finally speaks, he accepts that Gore's film made "a big difference" in alerting a huge segment of the public to the dangers of climate change. But there is a "but". "Part of my ambivalence is that he was Vice-President of the US at Kyoto and head of their delegation. The history books show that he was in Kyoto for seventeen hours. He just flew in, made a speech and effectively jetted out again. On his departure, he just told the US delegation 'show some flexibility'. . . . He delivered some awareness, but what else? I saw the film twice and there didn't seem to be anything there at all in terms of solutions. Solutions are half of Grian's mission statement."

Yet green activists are not beyond criticism. Some have been accused of overselling climate change and feeding grim stories to the media. Indeed, the head of the UN's Environment Programme, Achim Steiner, recently accepted that there was a tendency to overstate the risks in order to grab people's attention. Addressing a UN meeting in Valencia, he said the challenge now was to move on to solutions. Pat says he understands what Steiner was saying, but does not accept that the green lobby has been over-egging things. "Based on colleagues I know, I can't think of any occasion where anybody deliberately tried to be scary. Honest to God, I really can't. It's very different from answering a question from the media, and that being interpreted as being scary. If I'm scaring you, come and talk to me and I will present the evidence as to why I said that."

Presenting an argument is one thing, but getting it adopted as government policy is quite another. Since 1997, Pat has dealt with four Ministers for the Environment: Noel Dempsey, Martin Cullen, Dick Roche and, now, John Gormley. So how does he rate them, given his long-standing criticism of Ireland's political record on tackling climate change? He tells me Gormley isn't in office long enough for him to assess. "Under my watch, Dempsey was easily the best. He is considered a bit weird within FF . . . a little outside the tent. All ministers get the same briefing when they arrive in first and, out of the three,

Dempsey was the only one who listened to his climate people."

Yet the amount of "face-time" which Pat gets with ministers is limited in the extreme. "The only time I would have met any of those four ministers would be at the annual UN environmental meetings. Cullen and Roche didn't want to meet me, but still had to because I caught them. The rest of the time you are meeting, talking and lobbying to civil servants. They really do the work." I wonder if it was a bit like the BBC comedy *Yes Minister* in which the permanent civil servants held the real power? "They have huge power. No doubt about it."

Time and resources are continual problems, however. Put simply, it's very hard to track policy across eight departments as well as to secure funding. "Grian was set up very much as an experiment to see what we could do with no resources. Many years later, we still have not managed to get adequate resources." A lot of Grian's finance consequently comes from government grants, but that brings the headache of administration. "A large number of them require a detailed proposal first and a detailed analysis afterwards. It takes time and subtracts from your opportunities to make a difference in campaigning terms."

His workload increased still further in 2005 when he became a reviewer for the UN's Intergovernmental Panel on Climate Change, through his involvement with CAN. The IPCC produces a report, on average, every five years. In January 2006, Pat got his first draft – all 1,700 pages. "You have to be absolutely precise about any proposed changes. I was one of 600 people reviewing the document.

Governments were doing the same thing at the same time." Pat received a second draft in September 2006 and submitted his views again. Subsequently, the text of a summary for policy makers was sent to him in January 2007, with both documents being released finally that May.

A slight problem for Pat is that, prior to publication, government officials get the opportunity to edit the policy paper. Pat feels the IPCC therefore "gives more weight" to comments submitted from governments over reviewers like himself. He contends the IPCC is "more than a hostage to that fact". It's a moot point; however, the advantage of having all of the world's governments involved in the process is that they are bound by the final document. And even though many scientists felt the ultimate findings were watered down, there was still more than enough there to illustrate the scale of the problem posed by global warming.

Pat's area of speciality is on solutions: how can emissions be limited and preparations be made for the consequences of global warming? His suggestions are many and varied, but the key issue for him, above all others, is "getting a carbon price onto the street". This means that if the nations of the world sign up to a global agreement, and CO_2 is factored into all aspects of their economies, then "consumers who want to make a change are empowered". The flipside is that those who don't are hit punitively. This is why Pat spends so much time at international negotiations – to fight to secure a successor to the Kyoto Protocol which involves all nations and has at its heart a robust "carbon market". In his view, this is a "vital time" because the right deal will only be achieved if "people show up and make it want to happen".

As it stands, there's no guarantee this will happen. What's worrying him further is a gap between image and reality. For example, a British newspaper recently published a magazine supplement all about the environment and climate change solutions. Of ninety-six pages, thirteen featured advertisements for cars and three more on foreign holidays.

To operate in the way Pat does, therefore, takes both belief and determination. It also becomes totally time-consuming. Pat now finds relaxation "very difficult", given the overwhelming amount of work piling up on his desk – something he accepts is, at times, "sad". Somewhat unexpectedly, one place where he can turn off, to a degree, is sitting on a plane when returning from policy meetings. Maybe it's because he has only been flying a few years or because his mobile phone is off. Yet it's one of the few times that he can assess what has happened and evaluate what needs to be done while staring out the window. "It's kind of very strange, given aviation's effect on the climate. But, when you are flying back, you are exhausted; you are in this plane; you are over the clouds, which determine how we live; you are actually flying over the cause of climate change and participating in it. Yet I just find that's where I get into the zone."

He has become so deeply involved in the fight against climate change that he is beginning to forget what life was like before Grian was established. Indeed it's a campaign which, although he's a lifelong pacifist, he's beginning to see in military terms, particularly as the policy arguments intensify and the international meetings increase. "It's almost like a battle. This is what your whole training, for years before, was all about. You are now being flown in almost on a troop transporter. It's weird, but that's the way it's seemed to be recently. Every battle is different from the one previously, but all are part of a war."

7

20:20 Vision

It was two o'clock in the morning at the 1997 International Environment Conference in Kyoto, Japan. Noel Dempsey was ravenous. Why the canteen had closed at seven that evening, when international negotiations were going to go through the night, was a mystery to everyone. Leaving, even briefly, wasn't an option because this was the final stage of the talks process and EU environment ministers could be called into session at very short notice. "The negotiating team would come back every hour, bring us up to date and get a new mandate. So we had to hang around the whole time." The talks in question were among the most momentous international environmental negotiations ever to take place – finalising the Kyoto Protocol on reducing greenhouse gas emissions. "The ministerial involvement in the talks was to last three days, but it felt like six. The last twenty-four hours were very intense."

The Minister could function without sleep but not without food. "A few of us had missed meals earlier on due to talks. We couldn't stick it anymore and sent someone out to find the nearest take-away." A few minutes later and Noel Dempsey, along with eight or nine officials and Frank McDonald from *The Irish Times*, were tucking into a feast from McDonalds. Burgers and chips never tasted so good.

Six months earlier, and with the 1997 general election having gone Fianna Fáil's way, Noel Dempsey had been sitting at home in County Meath, wondering if he'd receive a call from the Taoiseach inviting him to join the Cabinet. "For three days before the announcement is made, it's very hard to focus on anything other than if you are in or out. It's actually very nerve-wracking." As he had been spokesman on the environment while in opposition, it was somewhat predictable that he would get the green brief. What did he think when it came? "I was delighted."

Though global warming is a big issue today, back in 1997 it was way down the priority list for both politicians and voters. "It wouldn't have been a burning issue for the election, I can tell you. Waste and planning were issues, but not climate change. Becoming a Minister and having to read up on the subject really brought it home to me." Effectively the new Minister had to take a climate change crash-course so he would be up to speed by the time he was in Japan. "It was fascinating in many ways – it was being thrown in at the deep end." Kyoto was a "huge learning experience", even if the process was "painful and painfully slow". The British Environment Minister, Michael Meacher, was leading the talks but "wasn't helped by his boss John Prescott, who didn't have the same grasp of it at all". In the end, a deal was done. "It was a long arduous thing. I think it was six o'clock in the morning when the Protocol was finally agreed."

The Protocol is important, from an international perspective, because it imposed the first legally binding restrictions on emissions of greenhouse gases – not on all signatories, but on developed countries like Ireland. Its importance is becoming increasingly striking in Ireland as our poor performance on limiting emissions begins to cost us more than a billion euro. It's also been something of a significant reference point for Noel Dempsey as he has changed briefs from Environment to Energy to Transport, with a detour at Education.

Noel Dempsey had got a taste for politics while studying at UCD and "made a conscious decision to pursue it" even before graduating. Why? "I liked it. It was a way of making a difference. I believe that, even though people think it's corny." He didn't have any family connections and so set out "very deliberately" to pursue it himself. The person who encouraged him to get involved further was Councillor John Bird and he was co-opted onto Meath County Council in 1977. Subsequent hard work clearly paid off: ten years later he was a TD; ten years after that, he was Minister for the Environment.

After Kyoto, Noel Dempsey was almost immediately back in the thick of more international negotiations. This time the deal under consideration was how much pain each country was going to have to take in order to ensure that the EU, as a bloc, reduced greenhouse gases by eight per cent by 2012. The Minister's mandate from cabinet was to sign up, as long as Ireland was allowed to increase its emissions by between fifteen and twenty per cent above 1990 levels. The figure was based on the argument that Ireland was just beginning to develop and should not be made to suffer as a result of the historic industrial output of Britain, Germany and France. The EU felt Ireland's target should be ten per cent.

Recalling the talks, Noel Dempsey says that the British Environment Minister Michael Meacher initially applied light pressure. "The argument he was using was that 'We're all in this together, friend. The EU has shown great leadership on this. It would be a terrible thing if the EU members couldn't agree among themselves.'" During the negotiations, each country had a meeting room and Meacher shuttled between them trying to secure agreement. "We would have had a negotiating team of three or four – the British had twenty." As the pressure ramped up, Meacher became more direct, using any and every weapon. "At one point Michael had a nice friendly chat with me, basically saying what a great job we were all doing in Northern Ireland and perhaps London could go a little bit further, if we were prepared to move now. It got down to that level of arm-twisting."

Once it approached endgame, Noel Dempsey decided to make a move. "I knew we wouldn't get away with fifteen per cent. It was a case of timing the concession." He waited until what appeared to be the second last round of talks. "I said to Michael Meacher, 'I will be killed for this but – if it helps you – bank on thirteen per cent. But I can't go any lower." Ireland was one of the first EU members to show a sign of compromise and a deal was done.

But Minister Dempsey then had to return to cabinet and, in a manner of speaking, explain himself. "I presented the report to government saying, 'This was our negotiating stance. We have got a very good deal at thirteen per cent above 1990 levels.' And I got cabinet endorsement. There wasn't too much difficulty getting that through cabinet at the end of the day." While he says there was "no great howl" about the deal from the opposition, they did point out that three other countries had used the same arguments as Ireland but secured larger emissions targets – Spain got fifteen per cent above 1990 levels, Greece got twenty-five per cent and Portugal got twenty-seven per cent.

The Minister concedes that part of the reason there wasn't a major row over the deal wasn't simply due to his fellow politicians agreeing with his analysis but also "because people didn't fully understand the implications of it". When it came to deciding on measures to achieve the target, things would change. At the time, however, he felt confident returning to his ministerial colleagues, having attended many cabinet meetings prior to his appointment. "I had been chief whip for two governments. I wasn't going to take anything from anyone."

While some ministers can be very stand-offish with the media, Noel Dempsey is the opposite. I can remember on several occasions talking about policy in his house in County Meath over tea and biscuits. He enjoys the nitty-gritty of political issues and doesn't try to avoid difficult topics. If you argue with him, he is ready with a counter-argument and will put it forcefully. Another aspect of his character is that he can be far more forthright than you'd expect which, on some occasions, must cause him trouble within the party.

After Kyoto and the EU talks, Minister Dempsey's next task was to map out how Ireland was going to achieve the stated aim of ensuring our greenhouse gas emissions didn't increase by more than thirteen per cent above 1990 levels. The roadmap was to be given the title the National Climate Change Strategy. His gambit was to get big business onboard by telling them it was better to get involved than to fight an impossible fight. "I said, 'Don't be dragged kicking and screaming into this. Let's look at the opportunities. Let's move forward and don't wait until the last minute.'"

Given the complications of such a policy document, the process was protracted and relied heavily on the civil servants. "It wasn't an easy task. We were trying to balance a whole range of things – economic growth, agricultural incomes, economic development, jobs and so on against environmental considerations. People generally didn't understand or didn't want to understand the implications." The policy dragged on for months. "There wouldn't have been a huge political engagement about it. Except when you came to cabinet. Then everyone politically had a reason not to sign up to the strategy." It was eventually published by the end of 2000.

It was clear for several years before the Climate Change Strategy was published that it was going to have to be robust if Ireland was to deliver on Kyoto. The economic think-tank, the ESRI, had predicted that Ireland's emissions would be twenty-eight per cent above 1990 levels, rather than thirteen per cent, if business-as-usual policies continued. What's interesting about this statistic is that it was based on lower levels of growth than

actually turned out to be the case. In the end, the two
measures in the Strategy which could be described as
being robust were the introduction of a green tax and
phasing out of coal-burning at the ESB's generation station
at Moneypoint.

As the Minister who published the strategy and spoke
publicly in favour of a green tax, why was it not possible
to introduce it? "I was pushing for a carbon levy right
through that period. It was resisted for five years, and
eventually was thrown out." I'm interested to hear if he
would say who was most opposed to the measure, but he
demurs on the grounds of cabinet confidentiality. Journalists
have long contended that the then Finance Minister, Charlie
McCreevy, wasn't impressed with the value of a green tax
and, in that, was strongly supported by Tánaiste Mary
Harney. Clearly the Taoiseach wasn't prepared to make it
a key priority for government.

The commitment to introduce a tax was never acted
upon. "It wasn't rejected on environmental grounds but
for economic reasons and because of powerful lobbying
by the business community." This separation of the environment from economic policies is crucial to understanding how
government thought at the time. The approach, in many
ways, continues today.

When the cabinet came to deal with the phasing-out of
coal-burning at Moneypoint it was clear to everyone that
coal produces lots of CO_2 and switching over to gas would
be much cleaner. Yet despite the clear environmental benefits,
the concept known as "security of supply" was deemed to
be of greater importance. "The view was that if you made
that change, you would increase the country's dependency

on gas. Then, if something went wrong, and supply was affected, you would be seriously compromised."

Opposition parties charged that by dropping a green tax and refusing to change over Moneypoint, the Climate Change Strategy had effectively been gutted. They argued that these decisions left the government with no option but to buy its way out of the problem. The way this is done is by the government purchasing "credits" for the amount of CO_2 Ireland has produced above its Kyoto limit. What does its chief architect think? "That would be unfair on the overall balance of things. It certainly didn't achieve everything that it was supposed to achieve. There was a lot achieved in agriculture – some by circumstances, some deliberate. On the Moneypoint issue, a reduction of its CO_2 emissions was achieved through the introduction of cleaner technology, but not as much as a complete conversion to gas would have."

The bottom line is that in January 2008, Ireland's greenhouse gas emissions were at twenty-four per cent above 1990 levels, and not the thirteen per cent to which Noel Dempsey signed up. Transport emissions are rising quickly. Low energy standards in houses are going to have a long legacy. The government argues strenuously that buying credits is part of the Kyoto Protocol and the €290 million allocated to date will be invested in reducing harmful emissions in developing countries around the world. However, purchasing credits was supposed to be a last resort. In Ireland's case, we simply didn't take the measures which would have reduced emissions.

However, journalists and green activists tend not to lay the blame on Noel Dempsey. While the Strategy clearly was not implemented in full, it's not because he didn't try hard enough. The media does not usually praise politicians, yet Frank McDonald of *The Irish Times* has described him as the best Minister for the Environment since the office was created – not just because of his push on climate change, but also due to the fact that he significantly reformed the planning system and introduced a policy platform for dealing with waste. Yet when reporters blame Fianna Fáil, he will stoutly defend the party and reject most of the criticism as "20:20 vision after the event".

After the 2002 general election, Minister Dempsey moved on to Education and Science, where he had quite a turbulent two years in office. Martin Cullen took over the reins at the Custom House (headquarters of the Department of the Environment). Noel Dempsey's policy was now being adapted, and on occasion changed, by his successor. On such occasions, does a former Minister bang the cabinet table and demand his previous strategy is carried through? "There's a fair degree of continuity of policy, but if the next Minister is not as convinced as you are, or maybe is not as strong in government as you are because you were around longer, then difficulties can arise."

By 2004, he was on the move again, this time to the very oddly combined Department of Communications, Energy and Marine. With his climate change experience, he decided to draw up Ireland's first energy policy for decades. "In some senses we had no energy policy. Des O'Malley was the last one to produce any kind of strategy, thirty years ago, and that was making the case for nuclear power." Once Dempsey took up the reigns of power, energy policy would begin to develop in two ways – assessing how to support new renewable energies while also carrying out reviews of organisations that already existed. "The ESB has dominated electricity in Ireland since the state was founded. It's done a great job and is absolutely pivotal to the success of the economy. But we needed to look beyond that."

Noel Dempsey's job was helped in no small fashion by the fact that climate change suddenly went mainstream. He describes it as an explosion in the public mind. The turning point, in his view, was the publication of the report on the financial impact of global warming by Sir Nicholas Stern in October 2006 and subsequent media coverage. The former economist from the World Bank concluded that the world faced a catastrophic tipping point within just a few years, unless it took financial decisions now to deal with climate change. "It was coming into that mainstream space. And then Stern just really confirmed it. I remember, in 2000 and 2001, wondering, 'How are we going to get people interested in climate change? How do we get the message across to people?' It's no trouble at all now."

The attempt to draft a White Paper on Energy was, once again, a convoluted process. There were no simple solutions. One such example which sticks in the Minister's mind was a Green Party campaign to retain the glass milk bottle. "I thought it was good thinking: washing and reusing the glass bottle. I wanted to do something about this until some genius came to me and said: 'Take the full environ-mental costs into account.' He produced this piece of paper which gave the comparative costs, down to milk bottles being much heavier than milk cartons and requiring more energy to deliver. When you took everything into account, the cartons were more environmentally friendly than the milk bottles."

When it came to energy, the big clash between environment and economic policy was the issue of burning peat to generate electricity. "If you were doing it purely on an environmental basis, you would not have peat in the energy mix." This is because bogs hold huge amounts of carbon, so when you drain them and burn peat, massive amounts of CO_2 are released into the atmosphere. But, once again, this wasn't considered to be serious enough to restrict the use of peat. Instead, given that Ireland is more than ninety per cent reliant on imported energy, like oil and gas, the fact that we had an available supply of native fuel was the deciding factor. "If we do not burn peat, Ireland would be even more dependent on fossil fuels. Peat does not account for a large percentage of our energy supply but, if there was a crisis or a strike, you would have supplies in reserve." Maintaining peat stations in the midlands was, to a certain extent, also social policy in that it provided employment in an area with few jobs on offer. How the new Taoiseach, Brian Cowen, will deal with this matter will be watched carefully, particularly as he hails from Offaly.

When the energy plan was finally published in March 2007, the most eye-catching commitment was the one which ensured that one-third of Ireland's electricity requirements should come from renewable sources, such as wind, by 2020. However, the one which caught Minister Dempsey's eye was wave power – something which he had seen personally at a testing site off Galway Bay. "I was a little sceptical at first, but by the time I came back in from the testing site I was convinced that it had, at least, big potential. After subsequent meetings, I became more and more convinced."

The opposition parties – including the Greens' Eamon Ryan – criticised the energy plan for having few deadlines and big issues such as transport and aviation appeared to have been left out. Another difficulty was that in the run-up to publication, the EU had announced that it was going to reduce emissions unilaterally by twenty per cent by 2020; and possibly by thirty per cent. Was his plan effectively out-of-date on the day it was published? "If you don't start somewhere and make a stab at it, nothing happens. We were conscious of what was happening EU-wide and where they were going. One of the stipulations in the energy policy is that it has to be reviewed on a regular basis."

While much of his time in office was spent on planning and policy matters, the Minister also tried to engage with the public through the Power of One campaign. Through ads on radio, TV and also newspapers, people were shown steps to reduce their energy. Did he follow this advice himself? "At home we switch off the lights and turn off the TV, rather than leaving it on standby. I'm using public transport and walking a bit more." However, he accepts he could do more. "I have to confess that the house isn't the most energy efficient. It's thirty-five years old and we've been talking for five years about whether to renovate or knock down part of it. Replacing the windows hasn't happened yet." His cabinet colleague Dermot Ahern is well known for his green credentials. What about the rest of them? "Certainly people are a lot more conscious, but I wasn't chasing after them around their department to see if they switched off the lights."

What makes Noel Dempsey of particular interest for climate change is that he has straddled the three key ministries of Environment, Energy and now Transport. His latest portfolio, as he accepts himself, is going to be the most difficult of all. However, he rejects criticism that the department has not adopted proper transport planning. "The department isn't going the wrong way – it's the public. Transport emissions are due to go up by 260 per cent by 2020. That can't be allowed to happen."

Blaming the public is easy, but Noel Dempsey was one of those sitting around the cabinet table as the commuter belt around Dublin exploded. Our high reliance on cars comes mainly from bad planning. So although we have a National Spatial Strategy, which provides a blueprint on how the country should develop sustainably, has it not failed spectacularly? "Obviously, the legacy we have is a result of not thinking forward. But the sprawl of Dublin didn't start in 2000. By the time the Plan was produced, most of the planning permissions that are being built now were already in place." The Minister's view is that it's too early to judge a twenty-year plan after five years. "The impact of the Plan won't be felt for fifteen or twenty years. It's not unfair to say there was terribly bad planning. But it's equally unfair to say the National Spatial Strategy has failed because of the sprawl of Dublin. It's from now on that there'll be no excuse." To that end, he's also introduced a bill to provide for the long overdue Dublin Transportation Authority which, it's claimed, will have powers to prevent development in areas which aren't serviced by public transport.

So what else is he going to do in transport? He will prioritise rail over roads but road-building will continue. "What some people fail to understand is that you are not going to be able to provide rail to every point in Ireland. There's not going to be a huge amount of extra rail, but we will improve what we have and improve the services. But people in rural Ireland need good quality roads." What about axing expensive projects like Dublin's Metro North in favour of building more light-rail systems which would carry more people? "I'm fed up with cheaper solutions. The cheaper solutions gave us the M50 with two lanes, and we are now spending a billion euro making it three lanes in both directions. I think that's exactly the wrong way to go."

Now we're on a roll: What about more buses or privatising some of the routes? Does it have to be 2012 before integrated ticketing is available across the board? How long does it take to roll out speed cameras?

But the Minister is back making plans. His latest is on sustainable transport and travel. "It's about the need to change our behaviour. It could be as simple as walking to school or cycling to work. Our responsibility is to try to provide the incentive, financial or not, for people to get out of cars or share their cars. What the action plan will do is spell out where we are, where we are going and what alternatives there are."

Given the scale of Ireland's transport problem, and how the brief has battered the reputations of Ministers

Mary O'Rourke, Seamus Brennan and Martin Cullen, does Dempsey think the position is a poisoned chalice? "Transport was a ministry I thought would be a good one. It can involve good news and have effects in your own constituency. A very big positive was that a €34 billion financial framework was in place." That, at least, was the feeling he had until he walked in to meet his senior civil servants for his first briefing. "That's when you get all the bad news. You go in full of joy and come out thinking, 'Why did I want this job?' I was delighted to get it, though."

Transport makes great copy for journalists like Treacy Hogan, Environment Correspondent of the *Irish Independent*, because there has been, repeatedly, a huge gap between big government plans and the reality on the ground. Noel Dempsey does not need any reminding about the extreme difficulty in achieving results and what it feels like to be in the spotlight. His recent and abrupt initiative to force learner drivers to be always accompanied by qualified motorists was a disaster. The spectre of an estimated 120,000 drivers being put off the road produced howls of outrage on RTÉ's *Liveline* and excoriating editorials. Other failed policy initiatives were dragged back into the limelight. A U-turn was quickly executed.

One new dynamic of this government is that it has two Green Party Ministers at the cabinet table. At the very least, it's an opportunity for a different type of discussion to take place. "Part of the difficulty that we've had up to now is that cost-benefit analysis does not factor in the environmental cost. It's coming but could take three or four years. I have to say that will be helped enormously by the fact that we have two people directly in government . . . I'm not saying that they are absolutely right, but that's where they're coming from, and I think it's important because we are too much the other way."

That's probably about as much as Noel Dempsey is ever going say about the mistakes of the past. If you ask a tough question, he will fight back. Take, for example, Ireland's poor performance regarding Kyoto. Analysts contend that either he cut a bad deal on Ireland's emissions target or the government he was part of totally failed to live up to its promises. "Either these people don't understand what we signed up to, or don't want to. We've to reach that target by 2012 and have four years left to run. People are deliberately trying convey the impression that we were supposed to achieve the targets now. We are not. Some people don't want to believe that the purchase of carbon is part of the deal." But he does concede that cutting emissions would have been preferable to purchasing credits. "I wouldn't say it's the ideal way of achieving the target by any manner of means."

With the departure of Bertie Ahern as Taoiseach, there was once again much focus on Noel Dempsey in April 2008, particularly when he was slower than his colleagues to offer his support for Brian Cowen. By his own admission, he's interested in the top job. "I have never denied that I had ambitions to go as far as possible in politics. Though I've never sat down and said, 'What do I have to do to get to the next stage?' I've always felt, 'to thine own self, be true'. You do your best and, if you're good enough, you will get there."

If climate change is the greatest challenge of this generation, and given Noel Dempsey's experience, he might not be too far away from achieving his aim. Then again, the average tenure of a Fianna Fáil leader is estimated to be 13.6 years. And with Brian Cowen having just taken over, there might not be an opening until 2022.

PART TWO

What's happening?

"This is the first age that's ever paid much attention to the future, which is a little ironic since we may not have one."

– Arthur C Clarke

8

Riding a Wave

Our addiction to oil, coal and gas is going to cost us big-time. Why? Because Ireland, according to the most recent statistics, is ninety-one per cent reliant on imported fuel, when the European average is just fifty per cent. This means that we're extremely vulnerable to increases in prices. In 2008, the price of a barrel of oil exceeded $100 for the first time, and then increased again. We can do little when these price spikes occur – other than pay more. It's not just big business which will be hit, but the ordinary consumers too.

A sobering statistic from Sustainable Energy Ireland suggests that our energy supply is set to grow by a further twenty-seven per cent between 2005 and 2020. The government response has been, in the first instance, to promote energy conservation. Yet it's only a matter of time before taxes are introduced which target polluting fuels, such as petrol, in order to make people think twice before using them. The ESB also has advanced plans to cut down on usage by charging more for people who use lots of electricity during peak hours. At the heart of these strategies is the principle that if you use fossil fuels, you pay.

Many experts believe things will get significantly worse due to the concept known as "peak oil". This is where demand overtakes supply, a problem compounded by the fact that it's going to become increasingly difficult, and expensive, to extract the remaining oil. What does that mean? Yes, consumers shell out more as oil prices shoot up.

One solution to the gloom is to reduce our energy imports, for example, by generating more electricity ourselves. The government wants one-third of our electricity to come from such green sources by 2020, which would not just reduce our importation of coal but significantly reduce our greenhouse gases. Wind energy is expected to provide the bulk of the power, but the emerging alternative source is wave energy.

It's not hard to see why, when you learn that Ireland's west coast is credited with the most energetic waves in the world, after their more than three thousand kilometre sweep across the Atlantic Ocean. Tapping this energy certainly looks possible, as studies have also concluded that Ireland has the most accessible wave resource in Europe. At the moment, though, it's still only at research stage.

A small group of pioneers is trying to turn experimental prototypes into machines that can be bought off the shelf and put to work. Two of them got involved for very different reasons: a Cork hotelier desperately wanted to get out of the business; while a physicist secured interesting results by playing with a tub of water at the back of his Carlow workshop. Mike Whelan and William Dick are very different types of character, but both have an entrepreneurial spirit which could result in the energy from waves being converted into electricity on a grand scale. If these pioneers can tap into this continuous flow of energy, we could significantly reduce our reliance on oil, gas and coal-fired power stations.

Mike Whelan comes originally from Ballycotton and now heads up a company called Ocean Energy. He's a ball of energy himself, speaking quickly and fluently about how he has ended up being involved in wave-power. He has always had an interest in the sea, having first started out as a diver working for oil companies in the Persian Gulf. His first job was a rescue operation after an explosion on an oil rig. "It was like the Piper Alpha disaster in the North Sea. Seventeen people killed. After that, I got offered a job in Dubai and worked there for a number of years. Mostly, I was either laying pipelines to bring oil ashore or building oil platforms. I didn't like drilling boreholes, as often you would end up sitting on a rig for a week doing nothing." It was a lucrative job and was followed by a well-paid tour of duty in the North Sea.

Mike returned to Ireland in 1975 and established a diving company which won contracts for work on the new Kinsale gas field. By 1980, he had diversified into tugboats. At times, this involved heading out into storms to rescue stricken vessels. He became keenly aware of the power of waves. "If you go into deep water, you get a long lazy swell which is easy to manage. When you come towards the shore, the wave has a lot more power. It happens at around a depth of thirty to forty metres – the seabed is rising, the wind is blowing and the power of the wave simply builds and builds." It's this zone which Whelan would later try to tap for electricity generation.

As soon as Mike would receive a phone call, the adrenaline would course through his body until the job was done. Small wonder, when you consider that he never turned back to harbour in any storm. "Going out in a force five, six or seven storm was common, but we would ride it out. Once a French vessel caught fire off the Wexford coast and the crew was picked up by another ship. When we left Cork harbour, the north-east winds freshened further. It was a sixteen-hour steam to the scene and, when we got there, the weather was absolutely horrendous. We had to stand by at the site for ten hours before the winds moderated."

William Dick is also a pioneer of wave technology and heads up a company called Wavebob. Like Mike, he has bounced around a variety of careers, starting out as a physicist working on the very early-stage computers. "You had to wind a handle to get them started. The big ones were like walking into a locker-room – metal doors with a fan on top." The first computer in the country was secured by Irish Sugar in 1958 to do basic accounting. The first scientific computer, an IBM 1620, was secured by Trinity College Dublin in 1962, the year William graduated. He began to work with a company in England called ICT. "I was what they called an electro-mechanical engineer, helping on computer design."

By 1968, he opted to return to Ireland to work with the drinks company C&C, installing their first computer and then becoming a senior manager. While he describes the job as fantastic, he decided to retire at the age of forty-two. Clearly the drive of the entrepreneur made him eschew the easy lifestyle. "I just held my nose and jumped. I didn't want to get stuck there, hanging on 'til my gold watch." It was a brave move, given that he had to leave his pension behind him too. "My younger brother Chris became chief medical officer at Guinness, got a lovely pension and is already fully retired. He's wondering what the hell I'm up to."

William is a carpenter by hobby and, sitting in his Carlow kitchen, I can see from the tables and counter-tops that he's highly skilled. In the early 1980s, the hobby became a job when he decided to get into the timber industry. The IDA was giving grants and a survey suggested there was a gap in the market at the value-added end. He saw his opportunity. "We bought three big kilns and put in a boiler fired on wood-waste – the first time in the country sawmill residue was used to heat kilns." Using bio-mass is now all the rage – back then it was an innovation. Technically, everything went well and they were open for business within fourteen months. The big problem was that, by 1982, the building industry had contracted and the market convulsed. "Timber from Canada was coming into Naas cheaper than we could get it off the Wicklow hills." The company was eventually sold and William set up a brokering business instead. However, he knew the market might turn. Fourteen months later he bought the company back, returned it to profit and sold it again at a much higher price.

He was learning hard business lessons which would be hugely beneficial when he would eventually get into the wave energy business. "I learned lessons you would never pick up in college. It was an interesting time – better than a Masters in Business Administration." Part of that on-the-job education was learning when and how to get tough. Getting cheques was one ongoing trial. "You were under pressure, there were bills to be paid and cheques were hard to get." So what was the secret, I wonder, of getting money from people with their hands firmly in their pockets? "When you're up against it, you say: 'We're going to pursue you in the courts and we're going to advertise that fact in the national newspapers in seven days' time'. It was just a bit of a gimmick. Usually there were lots of four-letter words but you'd get the cheque."

Mike Whelan was also trying to keep a business afloat in very tricky circumstances. There was a good financial return from salvage work but you had to be extremely careful about pricing. "You have to have some sort of a clause which ensures that if the weather deteriorates, and the cost rises, you are not left out of pocket. You are constantly asking yourself, 'Have I priced it right?'" By the 1990s, the market had shifted beyond his control. Firstly, it became cheaper to hire local drivers from other companies, rather than maintain a business. There was also vastly increased competition in the salvage and barge business. "I could see that I needed to make a big investment but I couldn't get a contract to warrant it." His solution? Get out and start all over again. "The place I ran my business from in Cobh was a superb quayside site, part of the old yacht club building. I decided to maximise its value and build a hotel." Construction work on The WatersEdge Hotel took a year and two weeks and Mike enjoyed every minute of it. But that's when the big problem started. "Six or eight months after the doors opened, I went completely crazy. Selling sausages and bottles of orange to kids, and then going off to clean a toilet bowl? I needed some form of diversification to retain my sanity."

The person he turned to was Dr Tony Lewis from the Hydraulics and Maritime Research Centre, or HMRC, at University College Cork. By chance, his plea for help was perfectly timed. "I knew Tony because we'd worked on submarine pipelines and cables all over the country." Mike's intention was to see what Tony knew about plans to site windmills offshore. In the event, a cup of coffee led to something quite different. "He said, 'Mike, we have a prototype trying to generate electricity from waves. We just can't seem to get it over a hump. We've been playing with the model in tanks but we're stuck.'"

Mike Whelan decided to investigate it further and have a look at the tiny prototype with big possibilities. "I had a look at what they were doing. I thought to myself instantly: 'Forget about the windmills, I could enjoy this.' I was hooked." Part of the appeal was that he felt his background made him uniquely qualified to push the development forward. "I knew they would need barges and mooring systems and I'd been up that road so many times. I had been dealing with heavy seas, trying to beat the tide and hang on to stuff in bad weather. I knew it."

William Dick's association with wave energy started at an earlier date than Mike Whelan's and was also far more gradual. After selling up his wood business, he got involved in land-use consultancy, mainly focusing on forestry, geology and water quality. "We showed that afforestation in certain regions in the west of Ireland, where you have lots of granite and no limestone, changes the surface-water acidity. We established that it means the eggs of salmon never hatch and you end up with rivers that are barren." A subsequent contract was for examining the possibility of growing crops, like willow, as a renewable fuel or biomass. This was the point where his life changed. "The people who were running this particular study, based in Enniskillen, asked me would I also do a chapter on the possibilities for tidal and wave power. I said: 'I don't know anything about tidal or wave power.' He said: 'Nobody does. So will you write the chapter?' It was intriguing."

William started to dig and uncovered that a lot of research was being conducted in Holland. A short phone call led to him catching a plane to meet two inventors working on a wave energy prototype. What was the meeting like? "Fascinating. Then, around 1997, the Irish government announced that there would be a competition for a wave energy device to be tested in Irish waters. The grant was £1 million. I made an application with the two Dutch guys but, unfortunately, we didn't win. There were only four or five applications and we were certainly one of the last two."

The rejection didn't depress William Dick. His response was to "scratch around" and try to come up with his own improved model. He focused on developing a prototype which would float on the sea and came up with an idea. "Think of a milk bottle. Turn it upside down in the sea, so there's a little air and a lot of water. Push that down with your finger and it will sink but then rise up again. It will have a natural frequency – it will bob." Applying his physics knowledge, he worked out how to time the bob to the natural pace of the waves which, in the case of the Atlantic, is ten seconds. The result was Wavebob, which is anchored by slack moorings so it can react to any wave size. Amazingly, he developed the initial stages himself. "I made my own models. It sounds crazy, but I tested the first prototypes in a tub of water in the back of the workshop." Within a short time he was also knocking on the door of Dr Tony Lewis at the HMRC for scientific testing in big tanks.

The next trick was to secure funding. At a wave energy conference in Athens, William met Mark White from the Marine Institute in Galway. "He said: 'You've got an idea – all right. But I'm telling you, good ideas won't work on their own. You are going to have to have some big strong players behind you.' He was right." William had been dealing with a Norwegian company about possible investment but discussions had become bogged down. In the event, the Marine Institute introduced William to the Belfast shipbuilders Harland and Wolff, a company also interested in offshore wind projects. On the third meeting, the owner, Fred Olsen, decided to drop in. "I had my early PowerPoint display ready and his officials said Mr Olsen will come in but, don't worry, he'll probably only stay five minutes. Well, he stayed two hours." William later went to Olsen's 16,000-acre estate in Scotland and his first offer was to invest £1 million in the company. Negotiations took off.

When Mike Whelan got involved in Ocean Energy, the prototype was only one-fiftieth the size of a full-scale model. "We played with it for a few years, until it looked as if it was really going to work. I literally stood in the tank holding on to this thing as Tony started up the wave machine." With decent results, they decided to build a larger model and put it through its paces in a bigger tank. After looking around Europe and as far away as Newfoundland, the team opted to go to Nantes, France, for a fortnight. "We went and had a look at their tank first. That determined the size of the model we would test. We didn't just build anything; we built the model to their scale."

They were taking no chances – all of the HMRC computers, cameras and scientific analysis gadgets were taken to France to ensure the results in Nantes wouldn't differ in any way from those secured in Cork. However, the journey over was a little less than scientific. "We loaded everything up into my jeep, with the model on the roof. We took the ferry from Ringaskiddy to Roscoff and then drove to Nantes with a big yellow thing on top of the jeep. It's a wonder we didn't cause any crashes on the way."

The Ocean Energy Buoy generates electricity in a different way to Wavebob. If you can imagine a six-metre wave hurtling towards the Irish coast, the buoy operates by trapping waves inside a chamber within its hull. The wave compresses air which is then forced through a turbine which rotates and generates electricity. "The key to its progress is simplicity – it has only one moving part. The only thing left is developing a hull that can withstand extreme weather conditions."

The results from France were "super". It was time to go to the next stage: building a twenty-eight-tonne quarter-sized prototype. The plan was greatly assisted by the fact that the Marine Institute had established a designated testing site about a kilometre and a half off Spiddal, in Galway Bay. "They told us the waves in the site were about a quarter the size of those in the Atlantic – so that determined the size of the hull. As it turned out, the waves were twenty per cent higher than the estimated six metres. We recorded one that was 8.2 metres."

Despite being battered by the sea for more than eight months, the buoy continued to function well. When the prototype was eventually towed back into dry-dock in Galway, it was covered in barnacles and mussels but was still in working order. Given that good results have continued to come, I ask Mike if he feels, at last, that he's cracked it? "We've met all our expectations and even gone beyond that. But, until we have a cable into the shore and we're lighting up so many homes, I'll hold back."

William Dick's company Wavebob had also secured great results and constructed its own quarter-sized prototype. Indeed, Galway Bay is the only place in the world where two such models, at such an advanced size, are being tested at the same time. I ask William if that means there is fierce rivalry between himself and Mike as they battle to the finishing line – a commercial full-scale unit which could be sold on demand. "No, no. We meet Mike Whelan regularly. There's obviously a rivalry, but I think it's good there's perceived to be competition. I think it's right that the government, Marine Institute and SEI are not backing just one player. May the best man win."

At one point in 2006, it looked as if William Dick's friendly battle with Mike Whelan could have been over due to financial problems arising from technical difficulties. William tells me: "It got very serious and very tough, but we survived. Some people are risk-adverse, some people get a buzz from taking risk."

Wavebob is back in serious pursuit of its dream again. William believes it's been helped, immeasurably, by the phenomenal €1.87 billion sale of the wind-energy company, Airtricity, which Eddie O'Connor established. "Airtricity has lifted the whole market up. We're not talking about €5 million or €10 million here, we're talking about something huge." The company also feels that, finally, the political masters are copping on to this potential. "The government were certainly not the most helpful in the early days. Yet Energy Minister Eamon Ryan, and Noel Dempsey before him, is now saying: 'We have to do this.' The Irish wave energy resource is hugely significant. The fact that technology is emerging here in Ireland, well, it couldn't emerge in a better place."

For Wavebob, the next stage is to build a full-scale prototype and put it in the water by 2009. It has received more financial backing – this time from the Swedish energy provider Vattenfall AB. "The prototype will probably be in Portugal because the wave climate is easier there. In the west of Ireland it can be savage. It's like skiing: you learn on the blue run, not on the black run." All things going well, the development stage of Wavebob should soon be complete. "I think we're transformed. We have a really good team."

With the finishing-line not too far off, I wonder if William Dick ever reflected on where his drive comes from. "The environmental issue is huge. Around 1973, I read a very interesting book, *Limits to Growth*, which was produced by the think-tank, Club of Rome. It focused on population growth, pollution and diminishing resources. If you were dismal, you would dig a hole and shoot yourself after reading it." But that isn't William Dick's nature – he's driven, not just in his commercial life but in his personal one too. "I've worked in China and Russia. I've sailed a yacht, with three others, to the high Arctic. I've landed on Rockall and sailed a small boat to the Azores." At his core, and despite the climate change doom and gloom, he's an optimist. "You have to be an optimist. While I'd be pretty pessimistic about human nature, there's a very good chance our attitude is changing. And that just might mean we can wise up."

The Ocean Energy Buoy is also heading towards full-scale production, according to Mike Whelan. "We're starting to look around at shipyards which would be able to build the full-scale prototype and we're having discussions with turbine manufacturers." Seeing as the first prototype could be carried on the top of his jeep, I ask him how big the final version would be. "The one we're going to build next will weigh 600 tonnes, and we'll be looking at putting two 750-kilowatt turbines on it which could generate enough electricity for more than 400 homes."

Given his cautious approach, does Mike Whelan think that this is overly optimistic? "No, but we will need an awful lot of help. It's not just financial, we'll need companies to produce the hulls and turbines. There I could see a problem – we still have not identified a turbine supplier who could supply the quantity we need. Up to this point, no-one has produced a commercial wave-energy unit. But I'm very optimistic it's going to happen."

With all emerging technologies, one has to be cautious. As it currently stands, no wave device is commercially available which can generate electricity from wave power. All we have are prototypes which have the potential to deliver, at some unknown point in the future. Financiers who are backing these pioneers are therefore taking a substantial risk.

However, it's unarguable that wave energy is one of the key opportunities for Ireland. The government is so certain it will happen that it has already begun to factor in how wave energy will contribute to Ireland's overall energy mix – experts think that the industry in Ireland could be worth as much as €2 billion by 2025. And there is no doubting the dedication with which Mike Whelan and William Dick are approaching the task. It's certainly a prize for whoever cracks it.

9

The Elephant in the Room

As the month of March approached every year, Ann Kehoe would start to experience sleepless nights. Lying in bed at her farmhouse down in the Suir valley, County Tipperary, the same phrase would roll through her mind: "My God, it's nearly lambing season." The month of March meant, in her words, "hard work". It's a reasonable description when you learn that she and her husband, Brian, tended a flock of 900 breeding ewes. Even more so, when you establish that they delivered all of the lambs with the assistance of just one friend, Christy. "I would get up at a quarter to six in the morning, as Brian would just be going to bed. We never left our sheep unattended while we were lambing. It was twenty-four hours a day and tremendously hard work. It's not that every sheep will need a helping hand. But the one you miss is the one that could be in difficulty. You have to be there."

Once a lamb is born, it's taken to a pen with the mother where they bond for forty-eight hours, before being released into the fields. Given the number of lambs and ewes, that involves hauling a lot of water and food to a lot of individual pens over a lot of days. It was in the middle of this intensive period a few years ago that Ann Kehoe realised something had to change. "One Sunday I lifted 106 buckets of water in over gates and into pens. I thought to myself, 'There has to be an easier way of making a living.'" It was something of a depressing conclusion, because both she and Brian were "very dedicated" to their flock. It wasn't the hard work on the farm which was weighing on their minds – it was the decreasing return for their labour. "It seemed no matter how we cut our costs, we always found our profit margin per acre to be so insignificant as to be non-existent."

Things crystallised for Ann when she came to draft a paper for the Irish Grassland Association on her plans for the farm for the following five years. "My conclusion was that if the price of lamb at least kept in line with the rate of inflation, then we would continue to do what we were doing. But if it didn't, we would have to consider cutting numbers and looking at other options." In the event, the prices slid. "Would you believe, in that first year, the average price of a lamb dropped by €4.75 to €72.55. That was it – we were finished. You could produce the quality but, at the end of the day, you were financially vulnerable."

The reason Ann had been invited to deliver her paper was that the financial relationship between farmers and the European Union was changing. "You used to need a sheep or a cow to draw down a subsidy. This has changed to a single-farm payment. It means that once you abide by certain conditions, you are paid an average of what you'd been given over a number of previous years." There was a sting in the tail, however, because the EU's agricultural supports are to be reviewed in 2013. "Nobody knows what's going to happen after 2013. In the meantime, the payment is going to be reduced by about four to five per cent every year."

For some farmers, the year-on-year decrease in supports is a major concern. However, for Ann and Brian, the change gave them the space to think about their future in farming. "We knew very quickly that this was going to give us an opportunity of a couple of years to put in place something which would guarantee us a good income from farming. This is what we really want to do." What was blindingly obvious to Ann Kehoe yet something nobody was really talking about – the elephant in the room, so to speak – was that energy crops were the coming industry. These are plants which can be converted after harvesting into biomass – in other words, pellets, chips or briquettes which can be burned for heat or power.

There are three main reasons why energy is also high on the political as well as the farming agenda. First, oil prices have jack-knifed and seem likely to stay above $100 a barrel; secondly, the demand for energy is increasing. Indeed, the International Energy Agency predicts that, without policy change, global energy demand is projected to increase by over fifty per cent between now and 2030. This means high fuel prices and, possibly, shortages. The third factor, of course, is climate change – because burning fossil fuels generates CO_2, which is increasingly going to cost big money.

The obvious way to ease these pressures is to have a domestic energy source with low or zero emissions. It's unsurprising, therefore, that the government has described renewable green energies as "critical" and "an integral part" of its plans. Its stated intention is for thirty-three per cent of our electricity to come from renewable sources by 2020, mostly from wind generation.

So where does biomass fit in? According to the government's plans, biomass would mainly be used to generate electricity. The White Paper states: "We are setting the target of thirty per cent [biomass] co-firing at the three state-owned peat power generation stations to be achieved progressively by 2015." It added, "We will . . . [also] set a target for biomass firing at Moneypoint generating station by 2010."

The cultivation of this biomass had a particular appeal to Ann. It wasn't just the excitement of getting into an emerging market; it also seemed to have removed a particularly frustrating component of the food industry – the middleman. "As sheep farmers, we have lost our way to the marketplace and no longer supply directly to the consumer. Somebody else does that for us; at times, there are three people between us and the end-user." The result of this situation is that, in her view, primary producers were unfairly squeezed financially. "The factories were, of course, driven by purchasing as cheaply as they could. It can be argued that's the marketplace. But can you imagine going into work every week and not knowing at the end what you were going to be paid?"

Getting involved in a new industry seemed to offer financial security. "When you investigate the renewable energy industry, you begin to get very excited about it. We knew this was our opportunity to find a way of efficiently producing a raw material, adding value to it and having nobody between us and the end-user. It's a lovely balance between supply and demand. It means people are getting a fair price for what they produce and the energy-user is getting a fair price for what they are purchasing. That's secure. That's sustainable. There's no disadvantage to it."

With a clear idea on the direction to go, the next step was to identify what would be the best crop to grow. She determined that it needed to be something which could grow efficiently and effectively while delivering the highest yield per acre on an annual basis. That crop, she decided, would be miscanthus. "It's a crop you put in the ground and it takes two years to establish. But after that, the crop grows twelve feet tall every year of its life; is harvested every year of its life; and does not require any herbicides, pesticides or artificial fertilisers. On harvesting it will yield seventeen to twenty tonnes of a biomass material, per hectare, every year for twenty years. If your circumstances change, and you need the land back, it's like taking out any crop."

In 2007, Ann and Brian moved from planning to action: they decided to plant thirty acres with miscanthus. "I just found it so exciting. Even when I look back on it now, I remember thinking: 'This is it, guys. This is the start of a new industry.'" They were so enthusiastic that they held an open day and up to forty other farmers turned up at the farm to have a look. The start, though, wasn't very auspicious. "I remember when the miscanthus rhizomes arrived, they were an ugly-looking mass of root. When they tipped the load, Brian said: 'Is that it?' We'd paid something like €24,000 for it." As well as not looking particularly pleasing to the eye, there are also some difficulties getting this root into the ground. "They are very difficult to plant evenly. It's not like seed that flows. You can't calibrate a drill and off you go." But planted it was, and Ann Kehoe has never looked back.

Miscanthus is a grass which originates in Asia and, with its liquefied stem, broadly resembles bamboo. In Ireland, it has become known as "elephant grass" but, as Ann quickly points out, this confuses it with a type of grass from Africa. She is adamant that miscanthus will be known as miscanthus but, as is the way of these things, the moniker "elephant grass" will probably stick.

For the general public, the logic of using these energy crops can initially be confusing. If burning matter produces harmful carbon dioxide, why is burning an energy crop considered to be environmentally sound? It's a question Ann is frequently asked as she tours the country promoting energy crops as a carbon-neutral alternative. "I think people have to think of it this way – if you grow something, it captures an amount of carbon. Therefore when you burn it, it only releases the carbon it has captured." This is the reason that biomass is considered to be "carbon neutral", unlike burning coal or oil, which only ever releases more CO_2 into the atmosphere. But, as Ann says, this is only part of the positive story. "Every time the plant grows, a lot of the carbon goes down into the root. As only what's above ground is burned, the miscanthus is actually carbon positive." Apart from taking more CO_2 out of the atmosphere than it puts in, another environmental advantage is that miscanthus does not require much nurturing.

Being involved in a commercial activity deemed to be "carbon positive" is a major plus because it has a value, which can be traded like a commodity. From the very outset, Ann Kehoe wanted to ensure that hard commerce was going to be at the core of her approach. If the new market didn't stand up financially on its own two feet, then it risked becoming dependent on supports. After her experience in the sheep industry, this was something she didn't want to happen, at any cost.

Before growing a single plant, Ann wanted to ensure she had fully investigated the market and had a detailed plan of how to press forward. The person she turned to, with just a blank piece of paper, was Gavin Maxwell, an international business development and bio-energy consultant from Coolfin Partnership, based in the south-east, who has developed and been an advisor on several UN projects. He examined the possibilities and had a clear view. "He opened up a whole new world for us as farmers. He was the one who said: 'Guys, you need to change the way you do business. Everything has to be based on commercial reality or don't do it. Take advantage of government grants but let them be a bonus. For God's sake, don't depend on those structures for your commercial success.' He was right." Gavin's other key recommendation was that a farmer shouldn't try to enter the biomass market as an individual. And so he began the work of designing and helping to develop a national framework which would, hopefully, reduce Ireland's dependence on imported fossil fuels and deliver positive effects for our energy security, economic stability and climate change.

The first step was to form the Green Energy Growers Association in 2005 and Ann has been its National Director ever since. Given that she'd just managed to escape from a time-intensive job, was this the smart thing to be doing? "I thought very carefully about this. I didn't underestimate for one minute the job we had to do or that it would involve going to a lot of meetings. But it was an incredible journey." Advertising the new organisation was going to be expensive so GEGA decided to go down the direct route. "We went to the ploughing championships and a couple of other big agricultural events with our stands. Bit by bit, we began to collect contact numbers of farmers who were really interested." Initially the biggest take-up was in the south-east, so they established their first company there. Each company has three sections. "One is where we produce our raw material, the biomass crop; the other is where it's processed; and the third is the services and marketing."

This business model was in place by 2007 when Ann and Brian planted their first thirty acres of miscanthus. However, as others in the area were also involved in the new company, the total amount of energy crop planted that year was over 440 acres. Even though some had been grown in Wexford the previous year, the new pioneers were a bit nervous. "Everyone kept going out saying: 'Where is it? Is this thing going to grow?' The next thing, little green shoots began to emerge. Everybody began to ring each other saying: 'I can see it!'"

Within a short time, it began to grow vigorously and was several feet tall, swaying in the wind. However, not everyone was convinced. "Our neighbours were still looking at us, thinking we were half mad. It's difficult to grow because it's hard to establish and there can be gaps in the crop. And no farmer wants that." But the confidence appears to have paid off. "We're going to quadruple in 2008 what we planted last year. And we will probably multiply it by ten the following year. That's the nature of farmers. They see something. Then they want to know how it's done. You have to absolutely demonstrate how you do it physically. Only then they get excited, and see the opportunities."

GEGA are not the only company getting involved in miscanthus. JHM Crops, based near Adare in County Limerick last year supplied a small trial of 100 tonnes of miscanthus to the Edenderry power station. In 2008, it's in partnership with Quinn's of Baltinglass, County Wicklow, to supply 1,000 tonnes. JHM Crops also supplies the domestic market with briquettes which ordinary householders can buy. The miscanthus briquettes light easily and burn well, but the heat value is a little less than a peat briquette. However, clearly there is no comparison when it comes to carbon footprints. JHM expects to supply 1,000 tonnes for the domestic market this year.

While GEGA was trying to convince farmers to move into the energy crop market with miscanthus, all of the media focus was on the possibilities of biofuels – like generating a diesel substitute fuel from oilseed rape. Why, I wonder, didn't they start growing oilseed rape, with its distinctive yellow flower, like everyone else? The possibility of getting immediate results was surely far more lucrative? "We spent twelve months doing serious research and looked very carefully at land use. Oilseed rape does not produce enough energy per acre and we decided we would end up using too much land." GEGA estimates that miscanthus has three-and-a-half times greater total energy yield than oilseed rape.

The concerns over making biofuel from crops like oilseed rape are getting bigger. The Department of Energy estimates that to substitute just two per cent of transport fuels here would end up using eighteen per cent of our tilled land. Given that our target for 2020 is to have a ten per cent substitution rate, it means we'll almost certainly have to import. This raises serious concerns because land in places like Brazil and Indonesia, which should be rainforest or used for food cultivation, could end up providing biofuels for Europe. The change in land-use from growing food to producing biofuels is said to be a contributory factor to a doubling of food prices over the past three years which, according to the World Bank, could push a hundred million people in poorer developing countries further into poverty.

Ann believes that, in future, woody biomass-type crops like miscanthus will become the most efficient way of producing the petrol substitute ethanol. Results from Sweden and the US are certainly encouraging and this is a boon to GEGA's plans. "Miscanthus is not our only crop, as you must have a mix. We are planting large-scale trial plots of industrial hemp and switchgrass. All three are woody biomass crops and can also be converted into ethanol. It's about research and making sure you have the roadmap going in the right direction."

There's a steely determination to Ann Kehoe. She dismisses the push towards oilseed rape as nothing more than "noise". It isn't going to deflect her from the plan.

"You get your head down. You focus on what needs to be done, and what's the most efficient way of doing it. You drive forward." She's also very direct and implementing the plan successfully is what she intends to do. "Anyone who knows where I come from would not be surprised that once I found the roadmap for producing energy profitably, I wouldn't fail. I'm tired, but I'm satisfied and confident that we're doing the right thing. We now have seven companies all working together in partnership to produce a secure, clean, long-term supply of energy."

But growing and processing biomass is one thing; finding someone to buy the product is quite another. How would she market the fuel? "People will be purchasing a carbon-neutral, high-spec, bio-energy fuel product." Things are complicated, however, because miscanthus does not yet have much of a track record in Ireland. This difficulty is something Ann is keenly aware of. "Take a large-scale hotel: if they want to convert from using oil or gas, they have to invest heavily in new technology. The first thing such a business looks at is, 'What's the payback time? And if I invest, am I guaranteed a secure supply of the fuel? Will they still be supplying me with high-quality fuel in five or twenty years' time?'" The twin strategy to alleviate those fears is to produce biomass on a national scale and always employ contracts of supply. It also works well for the farmers. "We don't ask anyone to grow a crop unless they have a contract of supply. That is the way business is done."

But biomass is an emerging industry and, as such, is not without its problems. For example, some people who have switched over to wood-pellet boilers have complained about getting access to adequate supplies. Increasing cost has also been an issue. Just as biofuels will probably have to be imported, it's contended that the country will end up importing pellets from abroad, which will cause more problems than it will solve. There have also been technical problems with wood-chips, including smoke escaping from some boilers. A major concern is the ability of many of the installers to actually do the job correctly. However, given that countries like Austria secure virtually all of their home-heating from biomass, these could well just be teething problems. The broad expectation is that this industry will take off. Certainly it's what Ann believes will happen. "Clear uncomplicated information and a high standard of customer service is what is missing from the marketplace. Our green energy service companies will provide this. It's going to happen."

With oil prices spiking, reducing energy payments by switching to renewable energy is certainly appealing to business. There is also an additional financial advantage in significantly reducing your carbon emissions through biomass. "From the day our crops are put into the ground, we have a registration system that tracks their carbon value, right through to the day they are sold. Because of the way they have been produced – regionally, with local processing and delivery – they will be a certified fossil fuel replacement." In essence, this means that the purchasing industry will have a carbon credit which, under the Kyoto Protocol, can be traded.

Given that biomass can reduce our reliance on expensive imported fuel as well as driving down dangerous emissions, Ann Kehoe can't fathom why the government does not institute a massive push. It seems to make total sense given Ireland's clear inability to keep its emissions within the target set by the Kyoto Protocol. She feels the government has opted for a "crazy, lazy and expensive" solution of buying its way out of the problem instead. "In the 2006 budget, the government set aside €270 million to purchase carbon credits to invest abroad, which will not solve any problem. At the same time, they set aside €14 million, over three years, to help establish energy crops. That's their priority. That's it in a nutshell. The €270 million is their talk. The €14 million is their action."

Despite government commitments in the White Paper on Energy, Ann's rather jaundiced view has been fixed after meeting ministers and senior officials from five departments. In her view, the work they put into lengthy submissions got little return. "When you're meeting them everything is fine and polite. Then you walk out the door and you don't hear from them again. It's really frustrating. To tell you the truth, you just have to get on with what you were doing – with or without them." The last Fianna Fáil/PD coalition government would hotly dispute such a claim but, for Ann Kehoe, it's just "green-wash". "There's been a shift in their language and, after that, nothing."

Her ire stems from a perception that the government is hindering rather than helping the biomass industry. She cites one example of an entrepreneur who received government grant-aid to build a pure plant oil facility, but was then turned down for excise relief, which made the business untenable.

The Department of Finance is a notoriously cautious beast. Fiscal policy usually moves slowly, with results expected from seed-money before larger tranches of investment can follow. It's a tried-and-tested approach, but Ann views it simply as wasting valuable time.

Another infuriating example for her is the Rural Environment Protection Scheme, or REPS, which is described by the Department of Agriculture as "a scheme designed to reward farmers [financially] for carrying out their farming activities in an environmentally friendly manner and to bring about environmental improvement on existing farms". Ann contends that it's simply not working. "You cannot plant any more than ten hectares of miscanthus, or willow, and be paid for REPS as well, even though it's a highly environmental plant." At this point, she pauses in exasperation before continuing: "They say one thing, then put obstacles in your way and don't listen when you tell them. But we will do it without them. One day they will have to turn to us."

Given that uphill struggle, I wonder why GEGA doesn't join forces with the Irish Farmers Association, but the idea of forging links with the representative body of 85,000 farm families is dismissed. "The IFA's tradition is food production. I also think they've concentrated so much on the EU and supports that they have failed to secure a safe place in the market. This is a new industry. The only thing this has in common with the food industry is that it needs land to produce its raw material." The IFA can point to numerous statements over many years in which it urged the government to take action in favour of the biomass industry, but Ann believes she has earned the right to be critical as a former very active member. "I was an elected member of their national council and I was also vice-chair of one of their national commodity committees." What about amalgamating with the Irish Bio-Energy Association in order to punch with more power? "We are the only organisation which is one hundred per cent focused on the production of energy from agriculture. We knew what we wanted to do. When you don't have to negotiate and compromise, you can stay focused."

Agriculture is the sector with Ireland's largest amount of greenhouse gas emissions – twenty-eight per cent of the total. In the main, they come from cattle burping methane after consuming grass. Yet, unlike transport, emissions in the agriculture sector are going down due to a reduction in cattle stock, through reform of the Common Agricultural Policy. In the latest statistics from the EPA, agriculture emissions dropped by 1.4 per cent in 2006. GEGA's firm belief is that biomass is the mechanism to accelerate and deepen that trend. However, it quotes analysts who contend that Ireland lags somewhere between six and ten years behind most of Europe. Ann is hopeful that new EU targets for reducing our emissions by twenty per cent by 2020 will focus minds. "Government will have to look at all sectors of the economy. But the one which has the absolute ability to create real emission reductions, right across the board, is agriculture." She also believes it offers a bright future to farmers who have become depressed at the gloomy outlook. "Our message is, let's do something for ourselves now; let's do it in a commercial way; let's own the process; and let's supply the customer."

This is a message which Ann Kehoe is taking right around the country on a weekly basis. As it is late February and we are talking in County Meath, I ask her if this means she is avoiding this year's lambing season? She laughs loudly. It turns out Brian is back in the Suir valley helping to deliver this year's lambs. The flock has been reduced to 200, from the original 900, and she is on her way back home. Brian had been on the phone earlier wanting to know when exactly she was going to get stuck in. Ann Kehoe may be focused on the future, but the past is not that far behind.

10

Dundalk to Dalymount, Daily

"Anyone I know who commutes from Dundalk to Dublin is shattered." Peter Halpin is one such commuter and has been doing it for the past year-and-a-half. "You work your backside off during the week: eat/work/sleep. You try to relax at the weekend only to prepare for the following week. The quality of life for people commuting is absolutely horrendous, in my honest opinion."

Horrendous it may be, but commuting by car is increasingly common right across the country. According to the latest data from the Central Statistics Office, the numbers of persons driving to work by car, lorry or van increased by 225,000 between 2002 and 2006 – a rise of twenty-two per cent. Of the 1.9 million workers in the state in April 2006, almost 1.1 million drive to work.

In Peter's case, hour after hour is spent sitting in his car going nowhere fast. "I've worked it out that it's fifteen hours a week – minimum – that I would spend in the car." One of the many sources of frustration is the snail-paced bumper-to-bumper traffic which he always hits at the city limits. "I live in Blackrock, which is just outside Dundalk. I found that from home to the Port Tunnel takes me a maximum of fifty minutes, with that fantastic stretch of motorway. However, the journey from Whitehall to Phibsborough will take the same amount of time."

One of the main reasons for the increase in car use, and the consequent traffic chaos, has been the explosion of housing development over the past decade. According to the Department of the Environment, 650,000 housing units were built in the past ten years. That's thirty-five per cent of the country's total housing stock. And if one thing defines this development, it is urban sprawl. A group of planners, known as Urban Forum, has concluded that the Greater Dublin Area is now approaching the size of Los Angeles but with only a quarter of the population. Because the density of this housing development has been so low, it makes these areas almost impossible to service by public transport.

So we all get into our cars. Peter says that boredom sets in quickly. "It's a case of turning on the radio and playing CDs to try to keep your mind active. I buy two or three CDs every month just to freshen things up. I change the stations – listen to a bit of music and then news. I'll make the odd phone call, with the hands-free set, just to pass the time. It can be quite challenging. The bumper-to-bumper traffic can be so irritating. You can't wind down the windows because of fumes coming out of the exhausts. I'm physically as well as mentally shattered most evenings when I go home, especially if I'm delayed. Your head is just melted due to the traffic."

Of course, this problem is not just about lifestyles or woeful planning but, more critically, climate change. Harmful emissions of carbon dioxide from transport are the fastest growing of any sector in Ireland. According to the Environmental Protection Agency, transport made up almost twenty per cent of all emissions in 2006. The Agency calculates that the rise of emissions in the transport sector alone has been 165 per cent since 1990. In shorthand, if we remain shackled to the car, it will choke us.

Peter Halpin is the commercial manager of Bohemians Football Club, which is currently based at Dalymount Park. Peter is a very open man but, when I inform him that I track the fortunes of Shamrock Rovers, he looks as if he might reconsider the interview. "My job is basically ensuring that money is coming into the club from sponsorship, advertising and pre-season friendly matches. Last year, we had Sunderland, which was brilliant – Roy Keane's first match as a manager." The game was a media and financial coup as the club netted lots of publicity and a full-house of 8,000 ensured a profit in excess of €100,000. So did he meet the Sunderland gaffer? "No. I was on my honeymoon. I organised the match and was then on a cruise off Miami. But I didn't mind. They invited me over in November for the derby game against Newcastle."

So if he lives near Dundalk and works in Dublin, is it not possible to take the train and reduce his carbon footprint? Like some commuters, Peter isn't enamoured of Iarnród Éireann. "A friend of mine works in Dundalk, but his wife-to-be travels up and down by train every day. She says she's lucky to get a seat. There are some days you may not get a seat at all. At half seven in the morning, if you're standing there going to work, it does not give you much motivation. That's especially so when you're tired and weary." The other problem for Peter is that, once he arrived at Connolly Station, he would have to get a taxi or bus to Dalymount Park. "You are still going into traffic, as we're not in the city centre."

Peter is honest enough to admit that "comfort" is a big issue when it comes to choosing to use his car. "The club has provided me with a five-door Seat Leon. I think it's most important to have comfort because you are sitting so long and it puts pressure on your back and legs. You need good support." There is one major pitfall of having a Bohemians company car. "It's plastered in club crests, which doesn't help in Dundalk." At home, he uses his wife's more discreet Volkswagen. The flipside of comfort is that you have to pay for it. "I'd put in roughly €80 a week, which is a lot of money for petrol. If I was out and about that could shoot up to €100."

If the private car is comfortable, easy-to-use and reliable, then it is incumbent on policy-makers to ensure that public transport tries to match it. Commuters will simply not get out of their cars otherwise and transport emissions will continue to soar. The government has a multi-billion-euro investment programme, Transport 21, but we are a long way behind our European neighbours. For example, in Luxembourg, which has an equally bad emissions record, every bus-stop has information on what time a bus is going to arrive. Here you get the fairly useless information on what time the bus left the terminus. Travel to San Francisco, and they've had integrated ticketing for decades which allows commuters to jump on and off trams and buses with no extra charge within a certain timeframe. Here, despite many, many promises, you have to pay for each journey. For many new estates in the commuter belts around our major cities, there is no adequate public transport at all. And so the car continues to reign.

Given that public transport solutions are some way off, Peter is looking at the other alternatives. One option which would be both environmentally friendly and a stress-buster is working from home, even for part of the week. "If I could transfer my job to home, it would make a hell of a difference to my quality of life." It's something he's already had a brief opportunity of trying out. "After Christmas I did one day at home a week. The difference was unbelievable. I did a Wednesday or a Thursday and it broke up the week brilliantly."

Peter feels that, apart from avoiding a three-hour commute, his work-rate improved. "I actually did a better job because I knew I had a bit of a break at home. It wasn't a day off – I had the laptop on and was taking the calls." It could be at least a partial solution to his problem. "It's definitely an option I'll be pursuing with my boss to see if it's possible. Even two days at home a week would be fantastic." Yet he's also around long enough to know that his employers will have the final say. "It depends on what the bosses think. You have to play by the rules."

If Peter can successfully work out an arrangement with his employers which allows him to work two days at home, his weekly transport carbon footprint would be cut by forty per cent. It's this type of change which is required across the board if the upward curve of transport emissions is going to be reversed.

At the moment, though, working from home is only a dream. And so petrol continues to be pumped into Peter's car, which is driven up and down the M1 between Dundalk and Dublin. What does he think about the CO_2 impact? "I wouldn't really know too much about the effect, but I know it's not good." Similarly, the broad issue of climate change is not yet really on his radar. "It's not really an issue for me and I guess I don't really take it into account." That's not to say that Peter doesn't want to know about it or to take action. For example, he and his wife "make a conscious effort to recycle as much as possible". Their house is three years old and the builder was clear that "insulation is essential".

In Peter's view, there's an information deficit on global warming and its impacts. "I get my information from whatever is in the newspapers but there's not been very much." He says attitudes have changed over the past two years – "people are realising that the environment is very important" – but the subject still does not have the profile of other problems. "It's a completely different issue with speeding and drink-driving. It's really in people's faces. Maybe an aggressive advertising campaign might highlight the problem of emissions." In April 2008, the government launched a climate change awareness campaign and, in Peter's view, it's clearly needed. "People might then sit up, listen and pay attention to what's happening. I guess it's all about getting the information but, up to now, it's not readily available."

Against that background, Peter continues to commute. Part of that reality is watching the clock in order to get out of the office before the evening rush. A few minutes can make all the difference. "If I take a shorter lunch and get away out of here at a quarter to five, I would be home in Dundalk at six o'clock. If I'm delayed at all, and leave after five, I'd be lucky to be at home in Dundalk by seven."

In the middle of drive-time chaos, the stress and tension mounts, particularly when other motorists start trying to cut in ahead of you. "Normally I wouldn't have a problem letting a car into the traffic. But if I am sitting there for fifteen minutes, it's a case of 'No way'. Let's be honest about it – why should I sit in traffic and they don't?" Sitting in his car every evening, going nowhere, he can see the stress on other drivers' faces. "It's just madness. I see people, even at traffic lights, and they're drifting away. In some ways, their life is drifting away too. They're staring around them or banging their hands on their head. Other people read newspapers. Maybe a car behind them will bang on the horn to make sure they drive on. It's just crazy."

The government's National Spatial Strategy was drafted in order to limit the voracious expansion of the capital and focus development in key centres around the country. It's a twenty-year plan which has been defined, in the first quarter, by massive development in the capital's commuter belt. Given that the building boom was not accompanied by public transport investment, it's not a surprise to learn that two of the top four ranked counties with households owning at least one car are Meath (ninety per cent) and Kildare (eighty-seven per cent). As house prices rocketed, many people moved even further afield, to counties such as Louth, Cavan, Laois and Wexford. This unsustainable urban sprawl is replicated around Cork, Limerick and Galway.

Sustainable development is usually defined as development that meets the needs of the present without compromising the next generation in meeting theirs. If that is to mean anything, the environment has to be integrated into every single aspect of policy. Some call this "climate proofing" – asking searching questions about how sustainable any development is going to be, before giving it the go-ahead: What impact will this development have on emissions? What role will public transport be able to play if a housing estate is granted planning permission? If this development takes place, what other pressure will there be for further development? Any fair analysis of development over the past ten years would have to say, from a sustainable development point of view, that it has failed the test badly.

The government will suggest that such an economic view is blinkered as it doesn't take account of the ending of emigration, the creation of full employment, the massive investment in public services and the reduction in national debt. On the question of the Spatial Strategy, it's argued that it's too early to judge its effectiveness. But having an integrated transport plan does not mean that development has to stop. Interestingly, while it's rare for a member of government to use the "f" word, in 2008 the Transport Minister, Noel Dempsey, accepted that transport plans had, to date, *failed* to deliver for the public.

The consequence of this failure is visible not only in increased car ownership and rising emissions but also in the pressurised life of commuters. For Peter, the combination of stress at work and exhaustion is beginning to weigh on him. It has made those close to him suggest that he needs to re-think what he's doing. "Friends, family, the wife – the whole lot. People can see it. Work can be pressurised, but the commute can add to that. For example, when you're nearly home and you get a call from someone who should have spoken to you that morning. When they're doing your head in over something silly and trivial, it can be very draining. You do question yourself – 'What am I at?'"

Many of his friends, who were in the same boat, have made changes. "One of my friends was commuting to Coolock – going up and down that road every day for ten years. Recently he's had back trouble – a slipped disc and two operations. I think a lot of it is attributed to driving non-stop." But luck has intervened and his employer has relocated to Dundalk. "He's now walking to work and going to the gym at half five. At that time in the old job, he wouldn't even be on the train heading home to Dundalk. He looks better, has lost weight and has fantastic energy levels."

With no sign of such a change coming Peter's way, I ask him: Why not simply move closer to your place of work? "I suppose it's easier said than done; we have a house in Blackrock with a mortgage. On top of that my wife has a permanent position and that's a fantastic card for her. It's the least attractive thing to do – move and leave family and friends behind. Maybe if you are nineteen or twenty you can uproot and move to London or wherever." The cost would also be a factor. "In Dublin we would pay at least €200,000 more for the equivalent of what we've bought in Blackrock. Buying houses there is extortionate." What about renting a place in Dublin? "I've worked hard enough not to throw away money on rent."

If he can't move to Dublin, why does he not change jobs and work in Dundalk? But that would probably mean getting out of football, a sport which has dominated his life. "I love football. I played up the North with a couple of teams and was then across the water on numerous trials with Crewe, Leicester and Wigan." However, injury cut his career short. "I got a bad back and an ankle injury. A few other things kicked in – my mother passed away. I just didn't have the get-up-and-go to get back at that level." To stay in the sport, he got into administration, working part-time with Dundalk United before taking up the premium job at Bohemians.

So if circumstance and bad government planning are preventing people from getting out of their polluting cars, what other solutions for the environment might be on the horizon? The option most people know about is bio-fuels. While products such as oilseed rape are now getting very bad press, there are also factories which are converting waste oils into diesel in a very efficient manner. Scientists are working on new forms of bio-fuels, such as deriving petrol from algae, which would negate the need for large tracts of land to grow crops.

The head of the ESB, Padraig McManus, has floated the idea that the government make plans to introduce electric cars. With increasing amounts of our electricity being generated from renewable sources, it could be a very clean way of solving part of the problem at least. In London, electric cars are being given priority in the city, including free parking and free battery charging. In Israel, where the main cities are within 150 kilometres of each other, Renault is planning to begin mass production of electric cars in 2011. Plans are being made to invest hundreds of millions of dollars in things like an extensive network of charging points and hundreds of stations where owners can replace empty batteries.

At home, the government is giving some financial incentives for hybrid cars, made by companies such as Toyota and Saab. In the United States, the focus for a long time has been on developing a car which runs on hydrogen and emits only water from its exhaust. However, it is not expected to become a viable option until 2020 at least.

The hard fact is that oil is going to continue to reign for quite some time. Which means that commuters like Peter have little option but to drive or make substantial lifestyle changes. And for him, it's about to get ten times more difficult. "My wife is expecting and the baby is due in June. With a family, it definitely has to take its toll. Whether it's a new baby or a ten-year-old, it has to take a wallop." He's beginning to look around at other people in the same situation. "My brother-in-law commutes from Drogheda to Dublin each week. He gets the train at eight in the morning and luckily only works a short distance from Connolly Station. However, he's home around half past six and gets to see just about twenty minutes of the kids before they go to bed. With that type of situation, the pressure on everyone at home is high."

What Peter's hoping for is that his luck will be in – that he can work at least one day from his home in Blackrock and that a plan by Bohemians to move their ground close to Dublin Airport will actually happen. Evening congestion would be minimal. "It's busy out as far as the Swords-Donabate turn-off. It's slow, slow, slow and then, suddenly, it's a case of: 'Where did the traffic go?' There's a small bottleneck at the toll plaza but, after that, it's free-wheel all the way home." Hope is what keeps him going. "Whatever has to be done – so be it, at the moment."

For the government, however, transport is going to remain a massive headache. Because the housing boom was allowed to let rip, we now have tens of thousands of houses built in areas which can't easily be serviced by public transport. This in turn means that people living in these commuter belts will continue to rely on cars. As a consequence, the possibility of reducing emissions seems remote. This will cause colossal problems because of the new EU target for Ireland – to reduce our national emissions by twenty per cent by the year 2020.

In the short term, the only expectation can be that the Celtic Tiger era of bad planning will mean we live with an emissions and lifestyle hangover which will cost the country, and its people, dearly. Rather than sustainable development, we've experienced only sustained development. And now it's time to pay the price.

Peter is facing the reality that his daily commute is going to cost him more too. It's almost certain now that a carbon tax will be introduced which will mean, at the very least, higher fuel prices. On top of that, it's also increasingly likely that the capital will have a congestion charge, like London, in which motorists are obliged to pay every time they enter the centre of the city. Once introduced, taxes only ever go one way – and that's up.

11

Forty Shades of Green

As RTÉ's Environment Correspondent, I often meet people on bright sunny days who quip: "This global warming is great!" Usually the joke is accompanied by something of a nervous giggle. Underneath the humour there's a realisation that our climate *is* changing and that can be a little unsettling. Trish Hegarty lives on Donegal's stunningly beautiful Inishowen peninsula and says that the weather has been "quite freaky" in recent years because storms have intensified. "It seems that every week there are really high winds. This winter was worse than last winter. Last winter was worse than the year before. The number of people around here who lost their trampolines was phenomenal! Ours ended up spread across three fields. We were lucky it didn't hit the car or go through a window in the house."

It is a sunny afternoon when I arrive at Trish's house at Shrove, which looks out over Lough Foyle and Rathlin

Island. With a cloudless blue sky, it's easy to make out the coast of Scotland in the distance. Inishowen is renowned for its magnificent beaches, which I can see now snaking around the shoreline. Yet, last summer, the golden sand on a blue-flag beach yielded further evidence of change. "I never saw so many brown jellyfish, the ones that sting, washed up dead all over the beach. I got stung only once by a jellyfish when I was young, and I was never out of the water. It's very unsettling."

Such changes may have been among the factors that caused people in Inishowen to examine what action they can take on the environment. Trish believes it has certainly focused minds. "I think it's contributed to a growing awareness of global warming. It's coming from everywhere – radio, television, newspapers and politicians – not just from the usual green sources."

Inishowen has started to look at the possibility of the peninsula being developed as a "green tourism" zone. The concept is to make the tourist business as eco-friendly as possible and thereby attract environmentally aware tourists. As head of Inis Communications, Trish Hegarty has been closely involved in how such a project might get off the ground.

Inishowen, Ireland's most northerly point, is about twenty-six miles in length and, at its greatest breadth, twenty-six miles across. The official website describes it as "undiscovered Ireland, a world apart, a timeless place". It's true. But there's a reason why things have not changed much here. "The obvious thing, when you look at it on the map, is that it's effectively cut off from the rest of Donegal by the border. It's so geographically isolated that it has suffered economically. In terms of development, particularly since the peace process, people have been trying to address some of those economic inequalities in the region."

Inishowen is right on the border and there's something of a constant crossover between Northern Ireland and the Republic. Leaving Derry City, you can find Northern-registered buses sporting advertisements for shopping centres in Letterkenny. A few minutes later, across the border, there are petrol stations with prices prominently displayed in sterling. These roads are fairly busy because, with relatively low employment levels in Inishowen, people are constantly on the move. "The main areas of employment used to be agriculture and fishing, but both have taken a hammering. There are people from all around here who travel to Dublin at the crack of dawn on Monday morning and only come back on Friday night." The fall-off in construction work will hit this area particularly hard. "There are areas in the Carndonagh–Malin district, where about sixty per cent of the employment would have been from construction. It's going to be scary."

One area of employment which continues to hold up is tourism. As in many coastal communities, visitors have increasingly opted to build. "It's been huge. Right from my window, I can see sixteen new houses in the surrounding twenty fields. About half of them are holiday homes." That said, it remains largely unspoilt and the strong feeling is that it's an area with a specific type of tourism potential. "There is an awareness across the industry that it's not a mass-tourism region and is never going to be. It does not have the huge water adventure centres or mass attractions.

But it does have Malin Head, beautiful scenery, unspoilt hills and beaches as well as Derry City and the airport on its doorstep. It's also still very friendly, due to a strong sense of identity forged from the historic isolation. People are conscious that if you are going to develop it, green tourism is the direction many people here, particularly in the B&B and self-catering sector, want to go."

Initial research was undertaken to find out what the green tourism market was like. Trish says the findings were very encouraging. "My colleague in the project, Dr Peter Doran, found that green tourists want to stay in smaller hotels or bed-and-breakfasts; they want to go to the countryside, and walk, cycle, fish and enjoy other outdoors activities." This was welcome news as Inishowen could offer all of this. More encouraging still was the positive profile of green tourists. "They come more often. They stay longer. They spend more money. They are worth going after."

I was interested to hear that B&Bs are a sought-after tourist option. For the past number of years, much of the media has been reporting that this sector has been shrinking; that people are leaving the industry in droves. What does Trish think? "It's a myth that the B&B is dead. It's only the Irish who think that. Seventy per cent of people who use B&Bs are international visitors. They still want what some of us are turning our noses up at: the genuine Irish family home experience. They are not necessarily going to get that in a big hotel, which may be the same as anywhere else in the world. They're looking for something that tells them they are in Ireland."

There are many B&Bs in Inishowen and the sector is undergoing a major transformation. "One of the hardest things for the B&B sector would have been coping with the need to do part of their business online. That's all changing. Town and Country Homes, which represents B&Bs, is about to launch a new website in which people can make live bookings. There's a huge amount of training and investment going on. They are also developing awareness of the niche markets."

According to Fáilte Ireland, there were 7.4 million overseas tourist visits to Ireland in 2006 which resulted in foreign exchange earnings of €4.7 billion and accounted for twelve per cent of jobs in the country. The data was released in 2007 as they launched an environmental action plan. A survey had revealed that eighty per cent of visitors rated Irish scenery as an important reason for visiting the country. The action plan was required because Fáilte Ireland felt "unprecedented economic growth in recent years has put the quality of this core tourism product under increasing pressure." For Inishowen, the statistics backed up the view that the market was worth going after and confirmed that the peninsula had what tourists clearly wanted. If it could be packaged in an eco-friendly way – all the better.

The next job was to assess what the Inishowen tourism sector would have to do before it could market itself as such a destination. It was clear the ground would have to be prepared very carefully. "You can't pretend that you're green, as that will only cause a backlash. You have got to look at all of the steps you have to take."

This research was made somewhat easier for Trish due to her background in journalism and experience in finding data from the relevant people. "I did a little bit of contract work at RTÉ initially and then spent many years at *The Irish Times*. I left Dublin for Derry next and worked for six years at BBC Radio Foyle. I loved broadcasting, but there comes a time when you have done all of the things you want to do. Also, working in news is not family-friendly. Shift work is terrible." She has now established her own PR consultancy, Inis Communications, which represents clients across the north west like Derry Port, and conducts media training and PR campaigns for clients such as Donegal County Council. But she also likes to work with community groups and the voluntary sector right on her doorstep. "I love Inishowen and have enjoyed working on a range of the tourism and community development PR projects."

With part-funding from the EU, the economic development agency Inishowen Rural Development Ltd. decided initially to focus on securing information about energy efficiency and using renewable energy resources. One of the first comprehensive steps was to conduct a survey of the tourism industry to find out about awareness. "We knew there was a general interest but didn't know how broad it was. The point of the survey was to find out what the average member of the industry felt – we didn't want to speak to the converted." The questionnaire yielded clear results – over two-thirds of those who responded said they did want to make changes but felt they didn't have the information they needed.

Encouragingly, the case for energy efficiency had already been won. Most said they had moved to energy-efficient light-bulbs and insulated boilers because energy costs were increasing. There was an obvious financial logic to it. Yet despite that interest, there was also a degree of financial scepticism about just what renewable energy could deliver: more than half of those surveyed were not sure there would be an economic return on an investment. "From a financial point of view, they were concerned about the risks. Now, that was despite grants being available. People do respond very quickly to incentives but, in this case, they still had serious concerns."

The twin fears of an absence of relevant information and questionable financial return were of note, particularly since the government's Power of One campaign was already up and running. The media is also regularly covering the issue. For Trish, there was a simple explanation. "There's a huge gap between what's happening at national level and what's taking place on the ground. It needs to be made locally relevant, to people and their businesses." But isn't all of the relevant information available for free on the web? "While some of it is excellent, not all of it is very accessible to the average consumer. Anyway, not everyone is computer-literate. Inishowen has amongst the poorest levels of educational attainment in the Republic. I don't think that you can call anything a mass information campaign, if it's only accessible online. It's a very narrow way to start out."

Another complication was that people in the Inishowen tourism business had been receiving mixed messages. The most illuminating example was the energy-efficient light-bulb. "In some B&Bs, they were being told by the Fáilte Ireland inspectors that they couldn't put them in, because the bulbs were not bright enough. So the B&Bs had gone out, bought all of these expensive light-bulbs and were then being told to take them all back out again. That puts people off. At any of the meetings or discussions we had, that was the subject which came up the most." Insulation was another problem area. "People were being told to put in double-glazing. But if you have a fantastic old country house, you are not going to want to rip out the original windows."

Armed with these insights, a plan was forged to move forward together. "IRDL felt that there was no point moving towards green tourism unless you are bringing the industry with you. We set up a body which was broadly representative of the entire tourism sector but also invited people with expertise on board." These included representatives from the renewable energy sector and Donegal County Council. Farmers were also asked to participate. "We did that because there's no point in trying to create a demand for renewable energy if you can't, for example, supply wood chips for the green boilers. So we wanted to address the chain of supply issue as well."

The quantum leap forward came in Spring 2007, when a six-week information campaign began in Inishowen which culminated in an energy fair, featuring exhibitions and workshops. The thirst for independent information from experts was already clear following an earlier half-day event. "There was a massive discussion. The event had already run over time, and I thought people would have headed for the door but they hung on as they really, really wanted to speak to an expert. There was a huge interest. People talked and talked."

The six-week campaign was supported by Donegal's Highland Radio, BBC Radio Foyle, Inishowen Community Radio and the local newspapers. Then Energy Minister Noel Dempsey also gave radio interviews on the first day.

The publicity was important because it had been "panic-stations" in the run-up to the big event. The problem was that it was extremely difficult to get details of the renewable energy industry in the region. "There was a huge database on the Sustainable Energy Ireland website, but it didn't have a regional listing and so you could have spent weeks trying to select companies." The crisis was solved when they secured a list from a neighbouring region who had run their own energy fair.

In the event, the Energy Fair was a huge success. "I'd spoken to people in much large urban areas and they were talking about similar events attracting a hundred people. By the time the doors had closed and the last person had gone, almost three hundred people had attended our fair. Inishowen is a small place with a population of around 32,000. All of those who exhibited said they got a huge response. Some of the companies got on to their offices during the day and said, 'Bring in more staff, we're swamped.' Many called afterwards and said, 'If you are doing another one, let us know.'"

Yet interest is one thing; taking out your wallet and handing over hard cash is quite another. Given the concern and scepticism which had been in evidence before, I want to know what uptake there had been in renewable energies following the fair. "Our impression is that people are going to be cautious. You can't really argue with that. If people rush in and buy something that does not work, it will be more detrimental in the longer term." That caution is also mirrored by the organisers. If members of the public liked a particular technology, they were being encouraged to talk to someone who had already installed it for a year. If they could, people were being advised to go and see it operating before considering a purchase.

In Trish's view, this caution is the crux of the problem – things would be moving far more quickly if the relevant information was tailored for, and delivered at, a local level. "It's very simple. It's not complex. A need for advice and information at a local level is what came back from the survey. It was great to have the fair because you could see wood-chip boilers and solar panels. It was amazing to walk around and see all the technology. But, where do you go then? You're not going to rely one hundred per cent on the person selling you the renewable energy. You want independent advice."

The experience of their European partners in Crete, Sweden and Germany, who were also running green tourism projects, was that providing local expertise or independent advice was the key to creating change successfully. "That's when people will start to buy – if they know that they can pick up the phone and talk to someone in their area. If it's someone in Dublin, who has never heard of where they come from, and is perhaps not interested in talking to them, it's unlikely to work."

The requirement to be seen to be acting in an environmentally friendly way is a challenge the entire Irish tourism industry faces. Fáilte Ireland has established an environmental policy and is actively seeking to help the accommodation sector get to grips with the new realities. Yet one area, known as the Greenbox, has already made major strides forward.

The Greenbox describes itself as "Ireland's first integrated ecotourism destination" and includes counties

Fermanagh and Leitrim as well as parts of Cavan, Sligo, Donegal and Monaghan. With a lot of focus on air miles and aviation pollution, Greenbox aims, in the main, to provide a holiday alternative. It describes ecotourism as "travel which is small-scale, low-impact, culturally sensitive, community-orientated, primarily nature-based, educational and capable of broadening people's minds and enlivening their souls." Like Inishowen, Greenbox has adopted a partnership model. But that is probably where the comparisons end. Greenbox is already significantly advanced and also has much wider aims. One of its key innovations has been to introduce the European eco-label, known as "the flower", to certify that tourist accommodation and services have a high environmental performance. To attain it, any B&B or hotel must complete a rigorous series of measures and then pass an inspection by independent testers.

According to Trish, the Inishowen team has been tracking their progress closely. "Andrew Ward is a great community activist in Inishowen, with strong leadership skills, and he's been very keen on the idea. What's been going on for the past five years here has effectively been small steps towards that." On a practical level, part of the reason for the small steps has been that any progress has usually been linked to the ability to secure government, agency or EU funding. Greenbox has, however, tried to give Inishowen some assistance. For example, some of the information leaflets now given to visitors to the peninsula result from receiving huge assistance from Greenbox. "We drew on the information which they had used, adapted pieces and then added some of our own. It's all very clear and straightforward: turn the room thermostat down; make sure you turn your television or DVD off; remember to switch your lights off; consider taking a shower rather than a bath. This lets the visitor know that Inishowen is thinking about the future."

That said, the aim of the Inishowen team is far more modest. For example, the process of qualification for an eco-flower is viewed as daunting. "It's an enormously demanding process. The manual is nearly two inches thick." A further hurdle is that a fee needs to be paid before the flower accreditation can be secured. It costs €300 to apply for the flower and then there's a minimum annual charge of €500. "In most other European countries, the cost of applying for an eco-flower was subsidised by central government. Here it wasn't. So there was already a financial deterrent for companies before they even entered the race."

If the Greenbox bar is, both financially and organisationally, too high, I want to know if it means the Inishowen green initiative is simply going to stall. "No. There are other forms of accreditation that you can go for. And it does not take away from the fact that we want to go in the direction of green tourism. The next phase will be when new EU funding programmes come in place. But outline plans for Inishowen are already highlighting energy projects and green tourism. IRDL, now the Inishowen Development Partnership, certainly intend keeping it on the agenda."

One of the big plans is the idea of turning one of the peninsula's villages into a self-sufficient zone. The big benefit would be that people could see renewable energies and environmental policies in action. This, it is hoped, would then encourage other individuals and businesses to follow suit. "It would take houses, B&Bs and small hotels through the process. That would show a path for everyone else." Rough proposals have been drawn up and the next step would be a feasibility study. Yet the central aim is, as it has always been, to be inclusive. "Andrew Ward is very clear that, rather than going for an ecotourism project, and only bringing a few people along, you are better off aiming for a simpler green tourism project and bringing everyone."

The only time Trish isn't fluid, forthright and forthcoming during our conversation is when I ask which village might be the leading contender to become the new sustainable centre of Inishowen. In fairness she clarifies that the proposal is only at outline plan at this stage and the location has not been formally discussed. However, she gives examples of the sorts of places where such a project might work: Ballyliffen, which has a number of hotels, tourist attractions and a track record of a cooperative approach and an enterprising attitude; another contender might be Greencastle, which has a tight-knit community and where fishermen are already working to convert waste-oil from the trawlers to fuel the community centre.

Whatever the finer details and hard choices, the overall message from Inishowen is that there's a confidence and desire to move forward and chase the emerging green tourism market. "People do want to make the change. It's not something that is beyond our reach." So what would make a difference, apart from the local approach, in getting people to switch over to renewable energies? In the end, it also comes down to money. "If you are running a small business like a B&B, there's no way you can afford to pay four to ten grand on installing renewable energy technology. You simply can't afford to do it. But if there was improved access to better grants then I think you would be more likely to. It could even be a loan. Other EU countries have shown that more generous grant aid will be taken up."

The fear in Inishowen is that the momentum and buzz could disappear; that the current sense of things being achievable could be lost. "I think at times it all seems insurmountable. Global warming is a huge thing – what can you do? As a whole, the level of change that needs to take place is overwhelming. So, when you can do small, practical things at local level, I believe it encourages people to think, 'Yes, I can make a difference.'"

12

Ice Ambassador

Lesley Butler stood on an Arctic peak and surveyed the sparkling white land below. "It was so quiet there. You felt like time itself had stopped. The Arctic is the most beautiful place in the world." The words which ran through her mind on that sunny day were tranquility, innocence and purity. It filled her with the feeling that "there was nothing to worry about". The extreme irony is that she had travelled to that white world to learn more about how global warming is threatening to eliminate Arctic sea-ice within just a few decades.

Six months after her Arctic adventure, Lesley Butler is still awe-struck by the experience. "We lived in snow caves. You have to find a specific area which doesn't have a ledge of snow over it where an avalanche might trap you. You would then dig right into it. You have to compact the snow above you constantly. We had special knives for cutting blocks out of the ice, which we'd use to build a wall around the cave. The lads were talking about how they did this all the time, in just a few hours if a storm was coming in. It took us a day. It was so hard physically. But they were really warm. You couldn't get me out of bed in the morning."

Just how did the life of an unassuming Westmeath postgraduate student change so dramatically that she ended up (briefly) living the life of an Arctic explorer? It all started in the cinema.

"I was actually on a date. I really wasn't interested in yer man, but I was very interested in the ad on the screen for the Climate Change College."

The college she is referring to was founded by ice cream manufacturer Ben and Jerry's. The ice cream giant began life as a hippy cottage industry in Vermont in the late 1970s and is now a global brand. In recent years, it has focused on global warming. The company asserts that its European production process is climate-neutral, from cow to cone, since April 2007. This is quite a feat given that cows are big generators of methane and ice cream requires a lot of cooling. One of its additional innovations has been to establish a college with the aim of "inspiring grassroots practical action on climate change". In Europe, one person is chosen from each of eight participating countries each year. Each lucky winner receives a €21,000 business mentoring programme, €7,000 monetary grant and, in 2007, a chance to participate in a fact-finding mission to the Arctic.

As she was studying sustainable development for her masters, Lesley already knew a little about the college. "It looked really good and I felt I'd have a good shot at it."

The financial package was certainly appealing, but the chance to travel to the top of the world was the real incentive. "The chance to actually get to see the Arctic was a big pull. It was the chance to travel somewhere you otherwise would never get the opportunity to go. I'd also get to see just how climate change was affecting the area. The statistics were already coming in that the Arctic could disappear in decades."

She followed up on the ad the next day and submitted an application, even though there were a few unanswered questions on the form. "I'm the kind of person who enters stuff and then completely forgets about it." However it must have been a fairly incisive proposal as she got a call to say that she'd made it to the last twenty-eight candidates. The next trick was to come up with a proposal for an eco-business. "They said: 'Lesley, how are you going to protect the environment?' So I started to write about how I was going to teach kids, but I then thought to myself: 'I know about environmental management. I really enjoy it. I actually one day want to open up my own business doing it. Why not do something like that?'"

The idea she hit upon was a company which would provide small and medium-sized businesses with the advice and means of becoming more environmentally friendly.

It's called EcoBiz. "My idea was there and I was really passionate about it. But they really helped me to structure it – to pinpoint my budget and set out whom I would contact."

The final part of the competition took place in Holland. "It was an intensive course oriented around finding out more about the participants. How did you work in a team? Were you a leader? Could you actually go with this idea? Can you see it through to the end?" There were three Irish finalists, including one of Lesley's best friends from college. I ask her if they tried to spike each other's projects. "No! It was grand. It was like a holiday. I was just happy to be in Holland."

A few weeks later, as she was sitting on the number 10 bus in Dublin her phone rang. A voice informed her that she'd just been chosen to travel to the Arctic as the Ben and Jerry's "climate change ambassador" for Ireland. "I burst out crying on the bus. Everyone was looking at me, but I didn't care."

She may have been bound for one of the most remote regions on the planet, but making contact that day with her parents in Mullingar proved problematic. "I had no credit for my mobile. So I had to run to the nearest shop, spend €20 and ring everybody in the phonebook." She then tried to grapple with the daily grind. "I went to the library in order to do my master's thesis." What do you know, little progress was made. "I just couldn't do it. I was so all over the place. I just knew that so many opportunities were going to be put in front of me. I was going to be able to help combat climate change and do it in a big way."

That was November 2006. By April, she was on her way to the high Arctic.

The whole expedition still feels like a dream. "Every single time I think about it, it feels like I'm reading a story about some lucky girl. I still can't believe that I actually went." The eight ambassadors first travelled to Norway where, in Tromso, she hooked up with a new boyfriend who, like she does, loves sardines. She was honest, though, and put it all up on her blog: "Okay, okay, he may be a bearded seal, but he was some kisser." Then it was off to the island of Svalbard, sometimes known as Spitzbergen. It's sited eighty-two degrees north, at the base of the Arctic, and was the location for a crash-course education on what was happening to the ice. "I was surrounded by really amazing people: guys who had climbed Mount Everest; Malaysian sky-divers; French scientists."

The aim of the exercise was to get up-close-and-personal with the impacts of rising temperatures. "It was a great chance to see how the Arctic was being affected by climate change. We were hearing from local people about how it had differed from just a few years ago. They were telling us about their fears for the future. For example, the people's lives have pivoted around snow and yet, soon, they are not even going to be able to go dog-sledding. It was an amazing opportunity."

The Arctic is currently a zone of global scientific focus – 2008 is International Polar Year, in which the ice caps are undergoing their greatest level of investigation in more than fifty years. It's hoped these studies will provide significant hard data to inform us, beyond any doubt, as to what's occurring. Some things are already known. Firstly, the Arctic is of particular importance to the global environment because its white blanket of snow and ice cover bounces back into space some of the heat coming from the sun. If the Arctic melts, the earth will heat up at a greater rate. Secondly, it's clear that, during the summer and autumn months, the Arctic sea-ice is melting at faster rates than before.

According to the UN, computer modelling suggests that large areas of the Arctic Ocean could lose year-round ice-cover by the end of this century. That report was released in February 2007, with a warning that temperatures at the poles were warming at twice the speed of the rest of the planet. No sooner had it been published than more ominous reports were circulating. Within two months, the US National Center for Atmospheric Research and the University of Colorado's National Snow and Ice Data Centre suggested the ice cover was retreating thirty years ahead of the UN's predictions. According to the World Wildlife Fund, which supports Ben and Jerry's climate change college: "Industrialised countries are carrying out an uncontrolled experiment [by continuing to emit large amounts of CO_2] and the Arctic is their first guinea pig. This is unethical and wrong. They must cut emissions of CO_2 now."

There were many things which stood out for Lesley about her Arctic adventure – the immense beauty of the scenery; the fact the sun was shining at half past eleven at night; but according to her blog, the best moment was the four-day trek into the wilderness. "It was a big challenge, travelling through knee-high snow pulling a thirty-kilogramme sleigh, cooking food and making water for nine starving people about four times a day and climbing to 600 metres on the steepest of mountains, but I managed to have a whale of a time. There really was no time to complain, to worry or back out. You were part of a team and things had to be done."

When she came back to Dublin she was "on a mission" to get things done. "I came back with the view, 'I've seen what's happening, this is very serious, so let's get it moving.'" Before leaving for the Arctic, Leslie had already been studying hard on climate change and its impacts. "It was all online. The college authorities would send us articles from the World Wildlife Fund, Ben and Jerry's and things like *An Inconvenient Truth*. It was only the second year of the climate change college, so it wasn't very structured, but it was great *craic*." There was also a conference in London which dealt with communications. "We have to try to convince people to change their lives and so that was a big part of it." The preparation meant that a lot was already in place when she came back from the Arctic. "EcoBiz went into full swing." The initial plan was to get thirty businesses in the Dublin region signed up to reduce their environmental impact over a year. Then EcoBiz would expand. "We had our first ten businesses lined up but we needed to get more. I held a concert in Dublin's Merrion Square. It got a lot of awareness going."

Even while attending Bolton Street for her masters degree, Lesley always felt she would end up in business. The management side of things particularly appealed to her – perhaps a result of her mother's commercial background. "My mother owned her own business and I was always oriented that way. Her business was furnishings, curtains, things like that." Her mother's business was also where Lesley first focused her environmental interest. "I used to think, if she just did this, she'd help protect the environment. If she got rid of her waste by recycling, rather than landfill, she'd help protect the environment and reduce her costs. It just clicked inside my head that this was the way to go."

Having just returned from a high-profile trip to the Arctic, Lesley found it easier than most to promote her business. She was on TV3, RTÉ, national and local newspapers. "The media really helped in promoting EcoBiz. It was straight into photo shoots – I felt like that Irish model, Glenda Gilson. It was really good." Yet the glamour was only ever going to last a short while. I ask her why she believes that her target clients, small and medium-sized businesses, will call her. From her answer –"Well they are not; I have to call them" – it is clear she knows that a hard road lies ahead. That said, she does receive some calls for assistance. "I often get contacted by the IT guy, or some clerk, who says, 'I don't think my business is very environmentally friendly. Will you come in and have a look?'"

EcoBiz has a website which lays out what services it offers. The company's slogan is "Improving the environment, improving your business". There are three stages: first an assessment with Lesley; this in turn leads to an action plan with goals; and finally assessment with, hopefully, the granting of a certificate. But why limit the client base to smaller companies? "I don't like to go to huge companies as they can manage it themselves. It's more the medium-sized businesses where there's a couple of staff and they want to develop."

The EcoBiz website lists the names of some of its clients; they include the specialist broadband provider, Centrecom. So what did EcoBiz do for them? Well, they drafted a letter to the owner of the building to ensure new environmental practices were adopted. The company also purchased a shredding machine whose output is recycled; filled the office with plants; and turned off all machines at night. They sound like small steps but, according to Lesley, they make a big difference to a company's environmental impact, to overall costs and to staff morale. "Top management always love any financial benefits as well as image improvement but the staff always like being part of something special. It shows that their company is concerned with the issue and they are involved."

Employee involvement is, indeed, one of the strategies which EcoBiz draws on. After an initial assessment, Lesley usually holds a meeting with the staff and recruits a "green team" to manage some of the changes. Usually it isn't difficult to get people involved. Indeed, Lesley believes any company which fails to deal with the environmental issue faces big trouble. "People are really putting up their hands now and saying: 'I'm part of a company which is not socially responsible.' Employees are saying: 'We don't want to be part of that'. It's becoming one of the reasons why people apply for a job with one company and not with another."

Another reason why employees seem to click with companies like EcoBiz is because the answers are often easy to understand and the tasks are manageable. "The three main areas are energy, waste and transport – so I ask: 'How are we going to minimise their impacts?' For example, computers can be such a waste of energy. If you are planning on purchasing, get a laptop instead. You can also get energy-saving modes; packages that turn the computers off at night. It's just that people are not aware. On waste, well done, people are recycling their paper. But you can also recycle your plastics, your organic waste . . . these days, you can pretty much get rid of all your waste through recycling and save money as well."

In hawking for business and analysing companies, one of the things which Lesley has identified is that many small and medium-sized companies tend to follow their competitors, when it comes to introducing green measures, rather than trail-blazing. "A lot of companies are unwilling to do anything more than recycle paper because it's not a common thing to do. I try to tell them, 'Just because the company down the road isn't installing energy-efficient light-bulbs does not mean that you shouldn't do it.' Hopefully these things will become common and standard in a few years' time."

Lesley Butler is without a doubt committed to her cause. Her business approach is to try constantly to secure a mind-shift within the company she's assessing. What she's not interested in is simply ensuring they've ticked off the boxes rather than undergoing fundamental change. "What I want to do centres on awareness. I don't want them to do this just because of legislation but because it saves money, helps protect the environment and improves the image. I want them to get the most benefits possible out of becoming environmentally friendly."

But, according to Lesley, there is still a bridge to be built between the theory and practice of "going green". Ask anyone in the street about the environment and they will tell you it's very important. But it's also quite possible they will have an energy-guzzling LCD-screen TV; a huge fridge, if not two; two cars; and take three foreign holidays a year. Is she pessimistic? In a word: no. "The Power of One campaign was really good at getting people to look at things again. It wasn't saying that we were all going to die from climate change but simply stating, 'Turn off your lights and save money.' Kids are learning so much about it all at school. They are going to go into the future knowing it."

Whatever about the children shaping the future, EcoBiz has to trade today. How does she get over a corporate fear-factor which must exist – let this woman into my company, and she'll run riot? "It is a big thing. People just don't know about the environmental path. So they are afraid I'll just tell them what to do and they will not know why they're doing it or if it'll mess up their business. But I show them how they can develop in a sustainable manner. If they are making changes, then why not do it this way? It's more environmentally friendly, will not increase your workload and will save money in the long run."

When she talks about her business, or even the trip to the Arctic, Lesley does not possess the intimidating zeal of some activists, which can so easily turn off ordinary people. While clearly clever and holding strong convictions, she's also light-hearted and self-deprecating. Maybe that comes from living with a group of people her own age who could elicit great fun from her if she gets too serious. But her belief, it seems, is contagious. "When we're sitting around, they'll say, 'We must get to the recycling centre and get rid of the stuff.' It's not that I prod them.

They could throw everything into one bin, but they don't. They separate it." She believes the awareness is out there so, even if mistakes are made, the positive thing is that people increasingly recognise what they should be doing. "I sometimes leave the lights on. But what's important is that I realise they shouldn't be on. Kids today have that. They realise, 'I must do that; I must close the door to keep the heat in.' That's sustainable. It's about preventing waste."

While praising the government for its energy-saving campaign, Lesley is also critical of its failure to incentivise businesses adequately to make the green change. "I think it can be made easier for business. I have cousins in Sweden who received a huge grant for putting solar panels on their house. The grant nearly covered the whole bill. It should be made easier for businesses to take on these environmental changes and get rewarded for it. If the Irish government made it really easy to change over, then companies and the public would do it. When I look at Ireland in comparison to other countries I think, 'God, we can do better.'"

There is no doubt that Ireland could do better in all areas on which EcoBiz focuses: waste, energy efficiency and transport. The question for Lesley Butler is whether EcoBiz can secure a foothold in a competitive market. For example, the Irish Small and Medium Enterprises Association, or ISME, already offers its members environmental advice. If needed, ISME refers people on to consultants who give reduced rates. The additional problem is that, in these more difficult economic times, owner-managers will want to know how quickly they will get a return for paying consultants like EcoBiz. As one man in the sector told me: "The environment is item fifteen on a list of sixteen." Lesley's focus on financial savings will help her pitch for business, but her prospective clients will want to see almost immediate results.

Already, Ben and Jerry's have moved on. The 2008 Irish climate change ambassador is Cara Augustenborg who recently completed her doctorate in Environmental Science and Engineering at the University of California and is now working as Intel's Research Fellow in the School of Business at Trinity College Dublin. Her business plan is to provide a unique service to Irish homeowners to help them do their own eco-renovation.

Just before finishing this book, I give Lesley a ring. Things have been tough but progress is being made as well. Her target for recruiting thirty businesses by November 2007 was missed, but nearly achieved by March 2008. The European Commission office in Ireland had signed up to EcoBiz and, as part of their arrangement, she had participated in a discussion with Environment Minister, John Gormley, on how to achieve new EU targets on cutting emissions by 2020. "It was a great experience." Lesley had also decided not to focus solely on long-term contracts but pick up "little jobs" like awareness meetings and lectures. The highlight, however, was that she'd been invited to work in Melbourne, Australia, for a number of weeks on an environmental project headed up by the company, Urban Trends.

Since that life-changing call on the No. 10 bus, things had at times been "over-whelming" and "extremely hard work" but also "a great laugh too". She remains highly motivated and has big plans for EcoBiz. And on days when the going is tough in the business world, she just thinks back to that Arctic adventure and reminds herself of the "most beautiful place in the world", which is increasingly under threat from a warming planet.

13

Back to School

Eoin Shimizu may be just eleven years old, but he is confidence personified. I am sitting in his primary school staff canteen in Galway city when he casually strolls in and introduces himself with an outstretched hand. He had been on TG4 a few days before, and is now being interviewed for a book, but it doesn't seem to take anything out of him. It isn't bravado – he is just a boy with a passion, a mission, and is quite comfortable with that.

Eoin attends Gaelscoil Dara, which is close to the city's Ffrench roundabout and the G Hotel. Once a TB ward, it became a school in 1984 and has more than 270 pupils, hailing from Clarinbridge, Oranmore, Moycullen and even as far off as the commuter-belt town of Gort. There are lots of Féile Náisiúnta plaques in a glass presentation box as you enter the green coloured single-story building. But, more importantly, outside on a prominent pole, hanging low on the windless day I visit, is a green flag.

When Eoin was eight, he was chosen to be the third-class representative on a new school committee aimed at securing the first green flag for his school. His teacher, Mari Ní Phuirséil, tells me he is definitely the most motivated boy in his class – the perfect candidate. His mother, Jane Shimizu, agrees, joking that he would also be the perfect candidate if you were looking for someone to lead the stand-on-your-head committee. She says that while Eoin has three sisters and one brother, he is the one who wants to get involved. "It's not just because he's the oldest; it's the personality. He's more interested in what's going on around him than the others. Eoin wants to know 'why'."

In order to secure their first green flag, the school's green committee was asked to focus initially on waste. Eoin picks up the story. "We went out after lunch-break and collected all the rubbish. We did a count of how many pieces of litter there were every day. At the end of the week, we put the total up on the notice board and then tried to see if we could reduce it the next time." The results can only be described as very impressive. "We reduced it from fifty-four bags of waste a week down to twelve."

I wonder if all of the waste was being generated by the school or if some of the rubbish was coming from elsewhere. Eoin tells me: "You could get very peculiar stuff – we found a nappy once." When I suggest that possibly one of his friends was hiding something, he throws his head back with an infectious laugh. Other material couldn't be blamed on students – a hub-cap and an exhaust pipe. In the main, though, it came from the school and the committee ensured that the volume of waste decreased significantly.

Making dramatic advances usually involves some pain, and it wasn't any different at Gaelscoil Dara. The big stumbling block was getting pupils to take home the waste which they brought with them to school. "We stopped my fellow students putting their lunch waste in the bins – yogurts, milk and wrappers." Did he require a big stick to enforce the law? "Picking up the litter was fine but stopping the lunch waste was harder – they weren't doing it. Eventually, after a couple of weeks, they did. We went around checking the bins to see how much rubbish there was."

The next step was to launch an assault on paper waste. Once again, significant results were secured. The total was reduced from nineteen bags of paper waste a week to just two. "We got bins for each class and you put the paper in there. You can put in plastics and other recycling things."

By the time Eoin had moved into fourth class, the green flag had yet to be secured but the campaign had gained momentum. "We put together a suggestion box seeking ideas, picked the best ones and started to try them." One of the top suggestions was to create a collage made totally from recyclable materials. "We made lots of art stuff from food wrappers and waste. We had to write out instructions. Everything was used. If you cut out a bit, the leftover piece had to go on something else."

That year, the school was informed that they were going to get a green flag as a reward for their efforts. A small number of children from the committee travelled to Dublin for the ceremony, and unfortunately Eoin was not one of them. However, the following Friday, a big celebration was held at the school. "The Mayor was here and we planted another tree. We all wore green. There was face-painting. A couple of green sweets as well!"

The green flag project is nothing short of a phenomenon. According to its website, over 2,700 primary, secondary and special schools are participating. That's sixty-five per cent of all schools in the country. The latest figures say that, so far, more than 1,100 have been awarded the green flag. The scheme is coordinated by An Taisce, in partnership with Local Authorities and supported financially by the Department of the Environment and several companies. It's not unique to Ireland and now operates in thirty-seven countries around the world, under the banner of "Eco-Schools".

To attain green flag status in Ireland, one must complete what is known as "the seven steps". The first thing to do is to form a green committee involving pupils, teachers and parents. After an environmental review, an action plan is drafted with targets and completion dates. It's a holistic approach with environmental issues being included in the curriculum and information meetings for parents and the wider community. If the school does attain a green flag, it has to be renewed every two years, which means the work is ongoing.

Gaelscoil Dara's involvement was, according to Mari, almost a natural follow-on to what was happening outside the classroom. "This was an up-and-coming thing all over the country. An awful lot of schools were endeavouring to make things better. From our point of view, our school wasn't the best at recycling. Yet when you went home, you were expected to do all of that. So to do that at home and have no choice about it, but then to come into school and sort of be able to forget about it, seemed really stupid."

Galway was probably Ireland's pioneering county when it came to segregating waste. Take the example of organic waste: in Dublin, a relatively small number of homes are involved in pilot projects, but the system has been in place for years in Galway for all households. I remember visiting the facility, which was basic but still impressive, at a time when a regular visitor was scaring off the scavenging birds – twice a month a bird of prey was brought in to give the gulls something to think about.

Jane Shimizu remembers that it was a fairly intensive learning curve. "We had to segregate our bins not only into green, brown and grey, but we also had to separate our magazines and papers from our plastics and from our tins."

When the project commenced in the school, the strategy was to get the children onboard first, rather than having the teachers telling them what to do. Eoin's teacher, Mari Ní Phuirséil, recalls it as a successful plan. "Some kids, like Eoin, were the first ones to really get motivated. They then started to motivate each other. What I've found over the past few years is that they go home and start motivating their parents. It spreads that way, far more quickly than if you had an adult or parent coming in and saying, 'This is what you have to do.'"

With the first green flag secured, Eoin stood down from the school's green committee. However, just one year later, he was drafted back on as the school tried to secure a second. He explains to me that, this time round, the emphasis is on energy conservation. "The focus is on electricity. We were told to switch off lights and computers, especially during break-time and PE. We put up signs and posters in the school." The green committee meets regularly and the minutes, with the progress achieved, are recorded for scrutiny later.

Eoin's mother, Jane, says it's becoming all-consuming. "I'll ask him, 'How was school?' 'Fine.' 'Did anything happen?' 'Oh yes! We had our green committee meeting!' 'Did anything else happen?' 'No.' He can't tell you if he learned anything or if the teacher was in or not. But he did have his committee meeting."

Parents are usually very supportive. The only sticking-point so far has been annoyance at food returning from the school. Jane does not have too much sympathy for these complaints. "If the children are not eating their yogurt, give it to them at home rather than getting the teacher to look after it." Mari is in agreement. "As a teacher, I have no qualms about sending stuff back home." Spillage from milk cartons is apparently a real parental bug-bear, but the policy isn't going to change.

On the positive side, parents and the wider community are being encouraged to take action through public meetings which are held at the school. Jane has found this both accessible and interesting. "At the one I attended, they were talking about wood-boilers and solar panels – what's in them; how you might go for a grant; where you source the wood pellets; how they get here. At the next one they might focus on wind farms." Mari attended the first meeting. "The message was, 'You will save a lot of money if you do these things.' I guess it was to give people the incentive."

But the green school project is not just about providing children and parents with information and hoping that they will change their ways. It's clear that when pupils get the message, they in turn start to apply pressure on the parents to take action. It's a process which Mari has been watching closely and the energy conservation campaign is a case in point. "I found the kids were saying, 'What can we do at home? Well, we don't have to leave every light in the house on. We can just have a lamp in the corner, rather than have the main light on all the time.' The kids are going home and giving out to mummy and daddy saying, 'You can switch the light off after you leave the room.'"

Eoin's mum has also watched this scenario unfold. She's in a good position too, as she, like Mari, is also a national school teacher (in Scoil Chaitríona Junior). "A child in fourth class came home and said to her parents that she had to read the ESB meter. Her father said: 'There's no key, it's locked, go back and tell your teacher.' A few days later she tells her parents: 'My teacher says you can get the keys for free in the ESB.' The pressure is 'nag, nag, nag, nag'. If the parents are not willing, but you keep sending homework like that, they'll have to do it."

Jane Shimizu is also associated with another school environmental project called "green wave". The aim is to get primary school kids to identify when the first buds open on trees and hedges. All the data they collect, from around the country, can assist in evaluating whether spring moves from south to north across the country or moves from the coast to the centre. From space, the budding season resembles a green wave moving across the planet. In years to come, it will also assist in determining whether the budding season is starting earlier each year if, as projected, temperatures keep rising. In 2007, more than 1,100 evaluations were submitted from schools to the organisers, Discover Science and Engineering.

Jane has found it to be a fantastic project for the children. "The first year there were just three species to identify: ash, hawthorn and horse-chestnut trees. You were to go out with your kids at least once a week and, nearer to the expected time, once a day. You have to record the very first day you see the buds and can clearly see the green. You record the information that day and send it off to them. The first time we sent it off by snail-mail. Last year, we did it through the computer."

What appeals to Jane is that it is clearly a scientific project. "Last year we had to look out for ash, hawthorn, horse-chestnut and primroses budding, as well as sightings of the first swallows. They are very precise. I noticed that one of our ash trees bloomed on the ninth of April. I sent it off to them saying our tree in school bloomed on the sixteenth but mine budded earlier. And I got an e-mail back saying, 'That's very unusual. We've asked our expert who wants to know if you are sure it's an ash tree and not a mountain ash. Could you send us a photo or a leaf from the tree?' It wasn't a case of slapping on any old data and sending it off."

The green school project has been an outstanding success across the country and, given that, maybe it's not surprising that there is ongoing pressure for it to be expanded. I ask Mari what type of assistance, either financial or moral, is forthcoming from the Department of Education. She pauses, smiles and says, "I don't think your tape is long enough. There's no motivation coming from the Department of Education. This project comes through An Taisce and the City Council. Any changes that were made here, were made on the ground and paid for by us." In Jane's view, the department's role is "miserable". Part of their ire stems from a sense that additional time-consuming work is being laid on their shoulders but, if the inspector arrived, they would be picked up for not working dutifully on core subjects.

It seems the schools work things out for themselves. Mari cites the efforts made to reduce water consumption as a good example. "Initially it was a cost-reduction thing. We had noticed that there appeared to be a huge wastage of water generally. So we started to monitor that. We got the plumbers in and shut off the water and tested if there were leakages. The next green flag will be all about water and we'll have it all done."

The possibility of having completed work in advance of the green flag assessment, raises the question as to whether, after three flags, there may be nothing left to do. Could the project close? Jane doesn't think so. "Our population changes all of the time. I'm in a junior school with just over 400 pupils. Once they get into third class, they move to another school. So our turnover is huge; every year we have 100 kids a year coming in and 100 kids leaving. So to do the same programme each time makes no difference, as you're hitting a new audience." Mari agrees: "There is no dead-end. Basically you are always going to have people to educate." Eoin chimes in: "You can't have a perfectly green anything."

By September 2008, Eoin will be attending secondary school. The plan is that his current green committee will link up with a similar group in his new school. But he's already looking beyond that – he hopes that there will be green flags for colleges too.

I am interested to find out if this focus on reducing waste, conserving energy and observing nature means that students are learning about climate change and its projected impacts. What does Eoin or his school friends know about it? "A lot of them are aware of global warming, maybe not to a very scientific extent. They are aware that something is going on." Some of the computer time is used by the kids to look up changes taking place around the world. Mari has noticed that the pupils have a particularly keen interest in disasters. "If they start seeing pictures of disastrous events, they say, 'This is amazing.' Doom and gloom. Ice melting. However, some of them also printed off complicated statistics on rising temperatures. We put it all up on the walls." Eoin liked a recent project by his friends on "The Sun". "It was a very good one. It showed the sun, and the waves coming out, and the gas levels, and how it'll get hotter on earth over the years." However, his particular favourite find was a funny cartoon about global warming causing sea levels to rise. "It was a cartoon with two sharks swimming around a tiny island with a man fishing. And one of the sharks says to the other, 'We'll only have to wait five more years to eat him!'"

Yet behind the fun and fascination, Mari believes that the children are recognising that climate change is a problem which is going to impact negatively on their future. She believes this is part of the motivation for them to push their parents into taking action now. Her belief is backed by a 2007 UN report for the Department of Education, which found that Irish fifteen-year-olds registered the highest level of awareness of environmental issues in the EU.

For Eoin, that connection is clear. "You are going to be going into the world which your past ancestors have ruined. That's not to say that all of them have been trying to ruin it, but without noticing it. Because you are going to get more floods, a lot more pollution and air contamination." If there were more like Eoin, maybe the outlook wouldn't be so bleak.

14

It Takes a Village

"A radical move but not rocket science" – that's how Gavin Harte describes the eco-village he's involved in developing in north County Tipperary. In an age when hype often does not match the reality, the Village project at Cloughjordan actually does live up to its billing. This is a plan devised by a community, rather than a property developer, to construct a village which has ecological, social and economic sustainability at its core. It couldn't be further away from the ugly developments of the past decade, in which mawkish suburban housing estates seem to fall out of the sky and land on unsuspecting towns and villages. This is very radical and could be a glimpse of a more environmentally friendly future.

Gavin Harte is well placed to talk about The Village, having been involved in the project for years. "It's a big whack of land – nearly seventy acres. It's going to be divided into three parts – the first will be planted up with trees, the second is going to be an active farm and the final third will be for housing development."

There will be 130 dwellings as well as community facilities and shops. In this case, the community has ignored the developers and taken the initiative – deciding through consensus how the future will look. "Fundamentally it's based on a cooperative housing company. It's not the usual system where the developer comes in, zips in the people and leaves. In this process, you have people saying things like: 'Well, I would like to see a tree circle.' The project is effectively developing at two levels. It's developing a physical infrastructure – roads and sewerage – but it's also developing a human infrastructure, a community. That said, it's a slower beast to move because of it."

Among the innovations is a street lighting system which is discreet enough to allow residents to see the sky. Rather than having individual boilers in each house, there will be a district heating system, based mainly on wood-chips but augmented by solar power. The car will be "a guest" rather than at the core of the development. Instead, priority is being given to a network of pathways and walkways. For Gavin, the boundary between houses is another key difference. "One of the ugliest sights I know of are estates where they have houses back-to-back, with a lattice of six-foot-high concrete block walls. It's such a barrier to getting to know your neighbours. Instead, ours is going to be a soft boundary with little trees and landscaping."

The financing of the project is broken into two parts. First, all stakeholders pay a sum for the communal infrastructure, like the heating, and outline planning permission. The average amount is €70,000; bigger houses can cost up to €140,000 while some of the smaller apartments can be as low as €40,000. Included in the charge is a share in the rest of the estate – including the pub. After that, individuals have to cover the building costs themselves. For Gavin, it makes sense. "Trying to finance the entire project was going to be a step too far. Anyhow, when it comes to building, everyone has a very personal atomised view as to how their home should look and feel."

The Village markets itself as a unique opportunity to have a superb quality of life. The trees are a good example of the approach, according to Gavin. "It's going to be an edible landscape. When we plant trees, they are not going to be torn down later. Many will bear fruit – pears and apples. When they come into season, it's going to be something beautiful." One third of the land area will be designated for farming and will include allotments for growing food as well as animal rearing. The emphasis, of course, is going to be on organic methods and best environmental practice.

So where did the idea come from? "The genesis of the project for me was our existing towns and villages. What makes them so wonderfully sustainable? They've been around for two or three hundred years, and there's a timeless quality about them. There is an economic system and social system which includes all functions of life. They are robust and solid. The Village model is effectively a small urban centre where people can live and enjoy all of the fundamental infrastructure of community, like libraries and halls. It just makes for better living." Gavin is modest and says the project isn't "trying to change the world" but he hopes that it will enable people to see just how the future could be.

Yet it has been a source of considerable annoyance to the activists that so much hard struggle has gone into getting things off the ground. "We didn't get Section 24 status or any of the other great tax-breaks available to mainstream developers. We get no credit for the fact that the project is trying to reduce its CO_2 footprint." Apparently the financial institutions took a little bit of time to get their collective heads around the project too. "We'd go to the bank and they'd say: 'What? 130 people wanting to secure mortgages from us for that?' They'd get in a flap and maybe it'd take six months to hammer out a unique deal. It's like struggling up a hill."

There have clearly been barriers to this sustainable village. At the source of it all, argues Gavin, is that people have grown so used to handing over decision-making to the builder. "To take that power back, and try to agree on a consensus . . . well, people are not very good at that anymore and have lost a lot of those skills."

The concept of sustainable communities and eco-villages is not new. Gavin first came across it in 1996. Now, in 2008, The Village is finally about to be constructed. He hopes it will lead to a change in the way development takes place in the country. Before he has time to elaborate, I ask him, whatever about hope, does he *believe* things will change? He laughs and considers the question for a moment. "If I was a developer, which would I feel better building? If I was still getting a good profit out of the build, but I knew that I was building something really positive for the community? I think that has to be an added benefit for that person. I think government has a function because people are often lazy – they won't do it unless there is legislation requiring them to. Look at the tribunals – the main root of all corruption has been the generation of enormous wealth from construction. Why did we get the planning so badly wrong in the past ten to fifteen years? Money."

I first met Gavin Harte in the late 1990s. He was one of the spokespeople for protestors trying to prevent Wicklow County Council from cutting down trees at the Glen of the Downs, in order to widen the road. The Council claimed that lives were at risk because the road had a capacity for 11,000 vehicles per day, but 25,000 were using it. It also continually pointed out that while 500 trees would have to be felled, 6,000 new trees were to be planted in the area.

Much to the Council's chagrin, its message was often lost when set against the dramatic images presented by the protestors. Influenced by the eco-warriors fighting a bypass at Newbury in Britain, they established an encampment at the Glen and then built a warren of tunnels and tree-top walkways to frustrate any chainsaw team which might attempt to cut down the trees, among which there were many very old oaks.

The protest, which lasted several years and involved multiple court cases, also had elements of high farce. On one occasion, it was reported a protestor used a bicycle clamp to secure his neck to the bumper of a Council lorry. When the Gardaí couldn't remove the lock, they took off the bumper, but the protestor remained attached to it. They then evaluated whether the now-arrested individual could be placed in the back of a squad car with the bumper around his neck. When this proved impossible, he was conveyed to Greystones Garda station on the back of a truck, still attached to a hunk of metal.

Gavin Harte was critical to ensuring the protest wasn't just dismissed as the actions of a lunatic fringe. While some of the protestors looked a bit "wild", Gavin was always well groomed. While some of the protestors used confrontational language, Gavin tailored his message to sound reasonable but very concerned. It wasn't that he was saying anything particularly different to the grizzled campaigners, but he had an ability to sell their message, bring an intellectual rigour and prevent it from being simply rubbished.

Nearly a decade on, Gavin stands over a campaign which ultimately failed in its primary objective – to save the old oak trees – but also cost the Council a lot of money. "I kind of knew that the campaign would lose. I knew the momentum that was at the Glen of the Downs. Yet, it was a place worthy of a stand – Ireland's first national nature reserve. It's such a valuable amenity and a natural barrier to the sprawl of Dublin. The city has now leaked through and is sprawling off down the east coast. But I'm very proud of that campaign. I think it was a very solid, robust and brave one in the light of the forces against us. What did it achieve? I think it saved about four metres on either side of the road."

Yet Gavin Harte doesn't restrict himself to explaining other people's actions to the media. He is also prepared to get on the wrong side of the law for what he views as the right cause. In 1999, he and six others were charged with forcible entry and damage to property at a County Wexford farm where genetically modified sugar beet was being grown by the Monsanto Corporation. Several dozen people had crawled under barbed wire and either pulled up the plants or walked all over them. Today, he is still a firm believer in direct action and civil disobedience. "Very often direct action is what's needed – and I get a kick out of it. It's definitely a young person's game though. All kids get a rewarding experience from being involved and government leaders need to hear the sound of protest. You shouldn't be afraid to join a march or sign a petition or get involved in direct action. Kids should be encouraged to do it."

If confrontation was something he was prepared to use as a tactic in the 1990s, he certainly re-evaluated the strategy after spending a year as national director of An Taisce, Ireland's National Trust and biggest environmental organisation. The white-hot row over one-off housing in the countryside is seared on his mind. In his view, An Taisce battled to stop urban sprawl and unsustainable development but was unfairly and unreasonably attacked. "It was an unhelpful debate. An Taisce was the scapegoat, portrayed for ten years as the anti-development bogeyman that would ruin small farmers. It was a very unfair assessment of an organisation which, in truth, was trying to do the right thing, with no resources from government. When it comes to development I was trying to show that there's an alternative to 'let's sprawl all over the country'. Unfortunately, everyone swallowed the bogeyman story, hook, line and sinker."

One thing that infuriated Harte during the long-running row was the portrayal of An Taisce members as serial objectors. "This was the phrase that was used so much: An Taisce objects to this, that and everything. The organisation had a statutory function to make observations on planning applications. I think it carried it out in a very positive way. It pointed out that one-off housing in the country has negative impacts. The whole argument got stuck in this emotional adversarial debate. It never really moved on, while the country was eaten up and swamped."

Reflecting on the row now, Harte says that what's disappointing is that there was rarely time to examine the areas of agreement between An Taisce and some of their chief opponents, like the Irish Rural Dwellers Association. Both would have agreed that our existing villages need support. "These are really positive examples of sustainable living, but we've let them fall into decline. So many are now like doughnuts – empty in the middle, with a ring of suburban housing a mile outside. You now have people commuting in and out of them. No one is really living above the shops any more. Any recent development has usually amounted to housing estates being slapped on."

Gavin worked well with the media during the Glen of the Downs protest; he knew something about journalism. He had previously worked on a magazine and also got used to drumming up press attention when trying to promote his rock band, Interference. In the early 1990s, he was one of the stars on an RTÉ TV programme called *Across the Line*, in which two young people were filmed trekking around the world on a financial shoe-string. He looked good, sounded smart and more presenting offers came his way. However, he turned them down. He would tell the *Sunday Independent* afterwards: "I found TV in general to be quite mad. I didn't like the idea of taking my brain and parking it outside the door for some director. I think I have a problem with authority. I don't really understand it."

Maybe partly due to that experience, but certainly because of his time with An Taisce, he's developed something of a jaundiced view of the media. He finds it infuriating that, when it comes to reporting on climate change, there appears to be a journalistic desire to always give equal weight to opposing views. "It's almost like we can't discuss climate change, unless we have someone who will talk against climate change. I know the media is trained in the concept of having both sides of the story, but you wouldn't give both sides of the story on the Holocaust in Germany. For some reason climate change seems to demand that we introduce the sceptics at all levels. It's like saying the world is flat."

In Ireland, there isn't a debate over whether humans are responsible for climate change, but there is a developing row over how quickly the country needs to change. Gavin Harte believes, given the scientific evidence, the media needs to convince the public of the importance of climate change and its role in tackling it. "Consistency of message is really important and the media need to take a lead on that. All sectors of society need to. It's like a war effort in a way. We've put so much time and effort into destroying the planet. If we took it on as an emergency, I think we could sort out some fundamental problems that have been dogging us and have a damn good time while we're at it. We could regenerate a new understanding about how we fit into this beautiful planet of ours, which has been so giving and generous. It would be a shame if our species missed that boat – and it is goodbye."

In April 2007, I clearly failed in what Gavin would regard as my responsibility. The occasion was the launch of a new umbrella organisation called Stop Climate Chaos. It was the first time major development groups such as Concern, Trócaire and Christian Aid joined a coalition alongside environmental groups like Friends of the Earth. On the morning before the media conference, I reported for RTÉ's *Morning Ireland* on the new organisation and its objectives, which included a demand that government commit itself to reducing greenhouse gas emissions by three per cent each year into the future. I concluded by saying that the public would be interested in hearing from the coalition about how this target could be achieved without having a significant detrimental impact on economic activity. Later when I walked into the media conference in Temple Bar, Gavin walked straight over to me: Why had I said that? Had anyone demanded that I say it? What's the agenda? He wasn't impressed by my reply that I raised the issue because I believed costing was an important component of the story for the public. Gavin wasn't aggressive in any shape or form, but it's an indication of the zeal and passion with which he approaches the subject. Others might well have felt the same that morning, but they certainly didn't tackle me.

Stop Climate Chaos is, however, a great indicator for Gavin Harte that things could be changing in a very positive way. It's an organisation with a broad inclusive agenda. Significantly, it has also shown an ability to motivate people – around 1,000 demonstrators marched through Dublin on a rainy December afternoon in 2007 in order to rally at the Custom House and call for more action on climate change. "I think what's happening in the non-governmental sector is that there is a commonality evolving, that is a planetary perspective. They are working together. You have environmental groups, development groups, youth groups, faith groups and business. I think in the next two years or so there will be a greater convergence."

So where did this desire for sustainable development come from? Gavin Harte is originally from Bandon, in west Cork. A heartland of Protestantism in the nineteenth century, it was described then as "Bandon – where even the pigs are Protestants". Gavin himself went to liberal Bandon Grammar School, which he describes as for "middle-class Protestants"; the comedian and presenter Graham Norton was a fellow pupil. Later Gavin studied at University College Cork where, surprise surprise, he became engrossed in student politics. But a feeling for the environment was with him from the very beginning, a sense he feels was imbued in him by his parents. "Scratch the surface of most people, and there's an inner yearning for open space and a beautiful view. Kids are perfect examples. For me personally, the connection with nature is something I feed from and feel better as a result of it. While I've involved myself in environmental campaigns, I look beyond them because it's also an economic and social issue. Sustainable development is the integrated and essential way of looking at things."

If there was a moment where this all crystallised, it was in 1991 when he read the seminal book by Dr James Lovelock, entitled *Gaia.* At its core, the work argued that the planet is a form of super-organism, a self-regulated living being. Lovelock is regarded by many as the main ideological leader of the global campaign to raise environmental awareness. However, he is also now something of a controversial individual because he argues that nuclear power can provide the only realistic energy solution which will save the planet from global warming. Whatever about the nuclear issue, the Gaia Theory made the concept of sustainable development a reality for Gavin Harte. "I think it was the first time I became aware of that global sense of climate. It explained an awful lot about how the planet works. It was clear you couldn't simply take a system out of context. You had to see the whole system in its unity. Lovelock was very clear in saying the climate of the planet is essentially the life-support system."

Today, Gavin Harte is setting up a sustainable development consultancy, and part of that work involves promoting and managing change. He understands the difficulties ahead in securing public buy-in to the massive changes which are going to be required to tackle climate change. Simply scaring people isn't going to work. "Switching off is a very natural response to bad news. If you look at alcohol addiction or drug dependency, anything which requires behavioural change, it's not an easy process. This is not a cake-walk. But there are tried-and-tested mechanisms out there. The trouble is, that requirement for change, as a response to global warming, is not being supported. I don't think fear will do it. I've never known people to do altruistic things through fear motivation. We are the first generation of our species to encounter this problem but we can get through it."

If that's to happen, the government is going to have to take the lead. On face value, Irish politics are also changing. In February 2007, Bertie Ahern promised a green revolution to members of Ógra Fianna Fáil in Galway, in advance of the election. By the end of the year, and with the election won, then Tánaiste Brian Cowen was saying the environment was going to be one of his top three priorities while serving as Minister for Finance. Gavin Harte has yet to be convinced. "Fianna Fáil calling for a green revolution is a very effective soundbite. If there is any meat behind it, I have not yet seen any evidence for it. We are still continuing with a massive National Development Plan which is based on fossil fuels. We are still proceeding down a route which presumes we are on a path of endless economic growth. All of the evidence is suggesting that the human footprint has exceeded the bio-carrying capacity of our planet. Unless we make a behavioural adjustment, and come back-on-track, we are on a fairly rocky road into the future. Bertie Ahern was right – a revolution is required. Whether Fianna Fáil will be a part of that revolution, well, I'm unconvinced."

Globally, it's common for our system of trade to operate without taking the environment into consideration. One of my recent favourites came in the form of a BBC report last year which showed prawns being collected off the shores of Scotland, loaded onto a plane, sent to Thailand for shelling, and then put back on a plane back to the UK for dipping and sale. It's an example which Gavin says "defies belief" but isn't that unusual. "The obvious explanation for the continuance of this eco-harmful trading system is that we have a fantastic resource in the form of fossil fuels. One gram of oil is equal to one man working for eight hours.

We have had this huge energy slave at our beck-and-call. But it's a party hang-over."

Whatever about his views on the wider climate change debate, I want to know how, as a pioneer of the eco-village concept, he would judge our current government as it grapples with sustainable development. Clearly the problems are immense – the country's five leading planning organisations maintain that Dublin will soon occupy the same surface area as Los Angeles, but with only a quarter of the population. For Gavin Harte, the Minister for the Environment, John Gormley, has to refocus on the village model. "Introducing guidelines for sustainable communities is something that could really have a positive impact on the planning process. The Village is a great example. There's no reason, in my mind, that councils couldn't do this more efficiently with more control and planning. Whenever I talk to people about The Village, they usually ask: 'What are your toilets going to be like?' But once you get over that, people just click with it. The Village needs to be held up as an attractive and beautiful place, with all of the social functions that are required. From a selling point-of-view, it's a no-brainer. The Minister needs to tackle it in that way."

PART THREE

What's going to happen?

"I always avoid prophesying beforehand because it is much better to prophesy after the event has already taken place."

– Winston Churchill

15

Model Man

John Sweeney believes there has been "a revolution in public consciousness" when it comes to understanding global warming and its implications. This enormous change has had direct consequences for him. "I'm somebody who worked in this area for thirty years. For twenty-five of them, I was a dusty academic that nobody wanted to know at all. Like many climate change scientists, we were thought of as working in an esoteric area which wasn't really relevant to society." After years of obscurity, however, John is stepping into the limelight. And that's because he is the leading expert on using computers to predict what impacts global warming will have on the country.

He is now in big demand from politicians, local authorities, business and agriculture. Why? Because anyone who is making plans for the future needs to know what the ramifications of climate change are going to be. The person who can best inform them is John Sweeney:

Professor, author of more than sixty scientific papers on climate change, head of the Irish Climate Analysis and Research Unit, contributing author and review editor of the UN's Intergovernmental Panel on Climate Change. The list goes on.

When you first look at John's predictions for changes in temperatures here, it all sounds like pretty good news. "By 2050 we expect to see summer temperatures across Ireland being 2.5 degrees warmer than they are today. That takes the midlands to a mean maximum, in June and July, of around twenty-three or twenty-four degrees centigrade. We expect the winter temperatures to have increased by a little less, maybe two degrees. That would mean the winters currently experienced in the south and west more or less being translated across the whole of the island."

Maybe we could do with an improvement in our sometimes dreary Irish weather. "We expect summer rainfall to be down by thirty or forty per cent in the south and east." But this is where the problems start. The biggest headache is going to be managing the supply of drinking water, something John predicts could cause a lot of tension. "For example, Dublin will have long since run out of water from the Blessington reservoir. It will have had to find a new source somewhere to the west. We expect to see competition for water among farmers, and those involved in tourism and recreation. That's going to be far more acute."

The beginnings of what is effectively about to become a resource war are already with us. Dublin City Council has been examining the logistics of tapping the Shannon.

However, farming, boating and angling organisations in the Shannon region are opposing the move because they contend it will damage their interests and the local environment. Why, they ask, should their resource be conveyed to the capital? If Dublin City Council fails to get its plans through, the fall-back is desalinisation – the conversion of sea-water into drinking water.

The significant reduction in summer rainfall will have obvious impacts on the country's rivers. "In the east and south, we expect to see more pollution problems if we continue to use the rivers for effluent disposal. There may only be half the volume of water in rivers that there once was. This means more fish-kills are likely and more in the way of pollutants due to lower oxygen levels."

A warmer climate would wreak havoc with agriculture. "Those dry summers will adversely affect Ireland's ability to grow grass. It will affect the ability to grow wheat and barley to some extent. It will certainly affect the productivity of cattle for milk and meat production." That's a particularly acute problem when you consider that the vast majority of the beef produced in Ireland is exported. "We expect to see changes in management; to see farmers using irrigation in places like Wexford, Carlow and even east Wicklow." The problem with that, as mentioned earlier, is that finding water could well be difficult and, with charges a distinct possibility, an expensive option too.

While the winters would be warmer, the weather would be more violent. "We also expect to see more hazards, in terms of a greater incidence of winter flooding, especially in the west. The once-in-a-century event, in some rivers, will have become the once-in-a-decade event. There will be a hazard to be managed and a risk to be taken into account." The damage could be dramatic, particularly around the Shannon estuary where the airport and several power stations would be at risk.

Mild winters would also lead to problems with pests. "Forestry may be afflicted by new diseases. These may well have to be taken into account due to the absence of frost during the winter and the associated problems of pests and diseases which can multiply far more rapidly in frost-free conditions." The recent mild winters are already being blamed for the explosion in rodent numbers – fewer icy nights equals far more rats.

But it wouldn't be all doom and gloom in 2050, according to the professor. "Interestingly, there may be benefits." Farmers, it seems will have opportunities. "It will certainly mean an ability to cultivate crops they can't grow at the moment. Crops like maize will become more widespread. We expect even the soya bean to begin making an appearance in the south."

There's also a strong chance that Ireland would be able to benefit from other nations' misery. "I think people will be coming here for cooler climes from unbearably hot places." In one fascinating prediction, he believes that the patterns of travel could be turned on their head. "Climate change could mean possibly even reversing the flow from the Mediterranean to Ireland during the summer." If it's too hot to travel abroad, then the Irish public may well choose to stay at home. "Domestic tourists would be spending more of their money here, causing maybe a rejuvenation of some of our traditional tourist towns of the south coast." If that's the case, it would seem logical to expect that many people will want to move here permanently.

Unlike less fortunate countries, such as Mali or Bangladesh, climate change would not turn Ireland into desert or submerge much of the land. But the impact would be overwhelmingly negative. "We may see habitats like the Burren threatened more by flooding in winter and drought in summer." Our bogs would be under threat, with one prediction suggesting they could be extinct in a generation. Human health would also be endangered. "We would have diseases related to heat. We may see increases in salmonella, as a result of those kind of heat-related multiplication of organisms. So we would have to be a lot more careful about food management and water cleanliness."

It's these types of predictions which have made Professor Sweeney a much sought-after commentator. Farmers want to know exactly what regions will be affected by drought and what areas will be hit by flooding. Industry wants to hear about which exports will expand and which will contract in the coming years. Politicians want to find out what lies ahead when they are drafting new policy. Local authorities want advice on where to locate flood barriers and suggestions on planning guidelines to take account of coastal erosion. Insurance companies are using his reports to bolster their demands for greater government investment in flood protection. According to the Irish Insurance Federation, they had to pay out €176 million after the last four floods. It describes as "paltry" the government allocation of €32 million for national flood defences in 2008, when the bill to protect Kilkenny alone would come to €48 million.

For someone who is so closely associated with climate change, John's first encounters with the subject came almost by accident. "I suppose I started researching climate change way back in the late 1970s. I did a big analysis of rainfall stations and found certain weather circulation types that were producing three or four times as much rain in Donegal as they were in Wexford. I started to examine how often this was happening. It occurred to me that the present climate was being influenced by what was happening in the longer term. I started looking again and that took me into questions of climate change."

John's particular area of expertise is in projecting what's going to happen in Ireland in several decades' time, with as much regional detail as possible. "People are almost fed up with the big global models. Really, what they need to know is: how will things change in my neck of the woods? If I'm a farmer in Wicklow or a fisherman in Donegal, what will it mean for me?" The way of answering these questions is to use computers – powerful computers which certainly were not available when he was studying for his doctorate at the University of Glasgow. "I worked with punch-cards which had holes in them. My supervisor's idea of sorting data then was sticking a big knitting needle through the holes and shaking it. That's how you did it in those days!"

Nowadays sophisticated number-crunching computer systems do the job. Two terms help explain how this works; the first is "climate modelling". This involves a computer programme developing a virtual representation of the earth and its weather systems. "We break up the surface of the earth into grid-squares. Usually it's 300 or 400 square kilometres in size." Then the scientists try to calculate what changes are likely to occur and what impacts they will have. "We try to evaluate the effect of a change, at any level, within that grid-square. There may be fifteen or twenty levels in the atmosphere and fifteen or twenty levels in the ocean. We then translate what effect that has on the next grid square, and on to the next grid square . . . to cover the whole world."

John accepts that climate modelling is "a very complicated exercise". Even with an advanced computer, capable of dealing with an immense amount of data, there are issues. "You have to process so many different calculations – millions. The effects of just one change may require a massive amount of computing. That means, even with the most complex and diverse models available today, and the most powerful computers we have today, we can't get to the kind of spatial resolution that we need for policy purposes." Put simply, the current grids cover a wide geographical area. While John can predict what's going to happen in the grid, he can't say accurately what will happen in a small part of it. For example , John can suggest what the future average rainfall will be across Ireland, but not what level of intensity Munster can expect.

This is where the second term comes into play: "downscaling". Scientists such as John use the data from the big grids and then extrapolate what these might mean for a smaller area by using regional computer modelling or statistical techniques. "It gives us a way of producing a confident prediction of how things change at a local level under present conditions, and also how things will change locally under future conditions. We test this out in the real world, in terms of the actual data available today."

People who are suspicious about the veracity of predicted climate change impacts usually hit out first at computer modelling and downscaling. One of the regular claims is that the computer programmers design climate models to give the answers they want. Another assertion is that the earth's atmosphere is so complex that it's impossible to include all of the meteorological possibilities. At the root of these suspicions is the belief that the predictions are bogus.

For John Sweeney, such concerns are totally misplaced because he contends the hard data backs up the computer predictions. "We only have to look at what's been happening in Ireland over the past forty years. We're seeing the changes in rainfall. We're seeing the changes in temperature. It's exactly as the models would have predicted. This would seem to give us confidence that they're working correctly, and they can be relied on for the future." Yet he's not over-selling the accuracy of the model predictions either. "While we're very confident that we have a very good handle on temperature, we are less confident about rainfall, as this can vary enormously. You can have a shower, and two or three miles away, it may never have rained all day."

While making that clear qualification, John is emphatic that the evidence which is being presented to policy-makers must be acted on urgently. "In my view, uncertainty is not at the level where we don't do anything about the problem. I think that this would be the biggest mistake for us to make – to use uncertainty as an excuse."

Part of the reason for this sense of urgency is that, even if all greenhouse gas emissions were stopped tomorrow, John says there's still a seventy per cent chance that the planet will hit the EU and UN threshold for dangerous climate change. This is where the planet heats up by two degrees above pre-industrial levels. "Some people have used the word 'tipping point'. It's maybe a bit emotive, but I think that it's nonetheless a value at which some parts of the earth's system may become a bit unstable. That's why the concerns are so heightened at the moment."

I wonder if dramatic changes in our climate could be just around the corner but, John says, this is not going to happen. "I think that it's fair to say that whether you adopt a very optimistic or pessimistic view of the future, the next thirty years will not be all that different. It will only be sixty or seventy years down the road that the choices we make today will be significant." This is why he passionately believes action needs to happen now. Failure by this generation to take action would have major consequences for the next generation. "We really have to start turning things around to avoid the hard choices which lie ahead."

Small wonder, therefore, that John is less than impressed by Irish government actions to date. What makes John stand out is that he has voiced his concerns. When I asked him about his choice of the word "delinquent" in a radio interview in early 2007 to describe government action on climate change, he quietly chuckled. Instead of it being a five-minute informative interview on RTÉ's *Morning*

Ireland, his employment of the "d-word" catapulted him into the headline news throughout the day.

The soft-spoken Scot does not shy away from tough-talking. He still thinks the term is appropriate. "Delinquent is a word which is quite apt for the government's performance on CO_2 emissions over the last decade. We've been one of the bad boys. A delinquent child is one who does not play by the rules, who does not obey authority, who does not do what is right because it's right. Those are things you could apply to Ireland."

Given that the 2007 general election was a matter of months away, John was certainly pushing the boat out by making such a forthright denunciation. It says something of his standing that, rather than being taken to task, Bertie Ahern invited him to address the Fianna Fáil Ard Fheis and made a point of attending his speech. At the same time, he was also relieved when John Gormley addressed the matter after taking over at the Custom House. "I was very reassured when the current Environment Minister was asked if the country was delinquent and he said 'yes'."

John remains forthright in his criticisms. He says that Ireland's failure to abide by its "generous" Kyoto Protocol emissions limit was a "misuse of the generosity" bestowed on the country by the EU. It was also wasteful. "The sooner you make a start, the easier it is to achieve things. But if you procrastinate for short-term gain, the pain will be greater down the road." The areas where we fall down include "unsustainable planning" in which "the housing boom was let rip"; "missing the boat" on sustainable transport; a National Spatial Strategy which was "mangled"; and commitments in the first National Climate Change Strategy which were never fulfilled. The net result is that the country has "lost a decade".

Maybe some of that no-nonsense approach comes from his upbringing in the disadvantaged Gorbals-Govan district of Glasgow. "It had a reputation for being tough, but it also had a reality of being a close-knit community. Tenement life had its advantages – there were twenty pairs of eyes watching you from various levels when you played. When I was being brought up there in the 1950s, it was a place of deprivation and poverty, but it was a happy place nonetheless."

It certainly didn't hold back his education; he qualified with a first-class degree from the University of Glasgow and then secured a PhD. He was a regular visitor to Ireland, as his mother had come from west Mayo and his father from north Donegal. Indeed, his initial visit to NUI Maynooth had more to do with his family than securing a lecturing post. "If I'm to be truthful, I guess I came to Maynooth as I saw it as an expenses-free trip to Mayo in 1978. I certainly never thought I'd end up here for thirty years." However, he describes himself as very happy there. It's a "delightful" and "rewarding" place to work where, because it's small, students and staff get to know each other very well.

As well as being a national figure, Professor Sweeney also plays an international role through his participation in the UN's Intergovernmental Panel on Climate Change, or IPCC – a position he secured after being nominated by the government. "I was a review editor on the agriculture chapter and made some contributions to the impacts chapter as well."

While the work was very rewarding, he was an uncomfortable witness to the interaction between science and politics – something which occurs when the IPCC reports every five years or so. The engagement happens in the week prior to the publication of the report, when the representatives of the world's governments effectively edit the final summary. It's easy to see why the scientists get tetchy, as they've been working on the text for five years. "The science is presented to the officials and they either buy into it or don't buy into it. If they buy into it, then they accept it. Obviously they then have a responsibility to act upon it."

The problem comes when the politicians flinch and edits are made. "It's quite depressing for very careful scientific work to be downplayed or dropped because a particular national government does not like it. National governments are like people and act in their own self-interest." I ask him for an example. "I recall a long discussion on the hazard posed by flooding to people living in the mega-deltas of southeast Asia. The scientific view was that hundreds of millions of people would be at risk. In the boiling-down process, it moved from hundreds to tens of millions." Why did he think that happened? "Particular governments may not wish to see something that stirs the blood too much back home." While he found it "very demoralising" to watch officials "deftly change meaning" and see emphasis "diluted" with "a comma here and a paragraph there", the "main thrust" of the scientific work was accepted and John remains a strong supporter of the work of the IPCC. He has little time for sceptics or suggestions that scientists are pressurised into saying global warming is manmade.

As the IPCC reports only every five years or so, John is refocusing on his domestic work, a lot of which is financially supported by the Environmental Protection Agency. His latest report, published in 2007, looked at what indicators currently exist to show that climate change is already occurring in Ireland. So far, he hasn't found anything which contradicts the modelling predictions. An analysis of meteorological records show that the highest rate of increase of temperature over a decade has occurred since 1980. While the warmest year on record was 1945, six of the ten warmest years have occurred since 1990. There has been a reduction in the number of frost days and a shortening of the frost season length. The annual rainfall has increased on the north and west coasts, with decreases or small increases in the south and east. The increases in intensity and frequency of extreme rainfall events provide a cause for concern as they may have a greater impact upon the environment, society and the economy. But more research is needed.

With the subject of climate change on everyone's lips, it has had a positive knock-on effect on Maynooth and his department. "It's true to say that there are a lot more people wanting to do research. There are also a lot more agencies inclined to fund research too. At the end of the 1990s, there was no funding for national environmental research. We either had European funds or nothing." Things have changed dramatically now. "I had one PhD student in 1999. I expect to have twenty next year alone." This, John argues, is good not just for his university but for the country as a whole. "It's very important that we build on that research capacity, because we're still a long way behind Europe. In Ireland, we've only been putting our eggs in two baskets: information technology and pharmaceuticals."

It would appear that the workload of Professor John Sweeney is only going to get heavier but it's something he's enjoying immensely. The fact that the country has finally "switched on" and become, in his word, "sensitised", is hugely encouraging. "I constantly get people writing and emailing me about dates of swallows arriving, frog-spawn hatching and changes in their local environment – especially from people of a more mature age. I think that's quite an interesting thing because you do need a long context. What I have learned is that scientists don't have all the answers. We sometimes pooh-pooh people's experiences of nature's workings, but I'm inclined to think we should listen to older people a lot too because they have that wealth of experience behind them. It may be that in twenty or thirty years' time we'll find there are grains of truth in their stories."

16

Spud

Who could have predicted that the humble potato could become the highest-profile victim of climate change in Ireland? Even though it is wrapped up with our cultural identity and deeply embedded in our collective psychology due to the Great Famine, growing potatoes here may soon be a thing of the past. The predicted warmer drier summers will mean that the spud could well become simply unviable in the east of the country. Put simply: no rain, no spuds.

However, for potato farmer David Rogers, from Ballyboughal in north Dublin, there is a more immediate threat – pasta. "For me, dinner isn't a dinner without the floury potato. But as I said to somebody: all our clients are going six foot under, and we're not getting the new ones."

He sees this change at his own table. "With my kids, it's tricky. When we were young there was nothing else and so we just ate the potatoes. The children now have a taste for pasta. They tell you they don't want potatoes or even like potatoes. Years ago, there was no choice."

David thinks all of the evidence is pointing to one conclusion – the potato will take a significant hit as times change. "Since the Celtic Tiger came ten or fifteen years ago, potato sales have been dropping at a fast rate. Consumption is definitely going down. People are moving to more convenient foods. At the best of times, it takes nearly an hour to peel and boil potatoes. With the next generation, potatoes might not feature at all."

Statistics from the website of the Irish Farmers' Association would seem to bear that out. Between 1980 and 2005, the amount of land used for growing potatoes has reduced from 45,000 hectares to 10,000 hectares; the volume of harvested potatoes has dropped from one million tonnes to 350,000; and the number of commercial potato growers has shrunk from 1,600 to 600. Could it be that consumer choice may kill off the potato before the climate gets the chance to?

Would you believe, 2008 has been designated by the United Nations as International Year of the Potato. It sounds like an extraordinary award for the modest spud, but the experts contend that the potato is the third most important food crop in the world. It is lauded for being capable of reducing hunger and poverty, while improving food security for millions. That's because as little as one hundred grams supplies about ten per cent of the recommended daily allowance of protein for children. It also supplies thiamin, niacin, vitamin B6, vitamin C and folic acid.

The origins of the potato go back to South America where it was cultivated extensively for thousands of years. Europeans came into contact with the spud as late as the 1500s, when the Spanish conquistadors continued their blood-soaked tramp through Peru on an insatiable hunt for gold. Some believe it was first brought to Youghal, County Cork by Sir Walter Raleigh. Whatever about the accuracy of that contention, it wasn't generally grown here for another 200 years.

The crop proved extremely popular because of its ability to grow in a variety of soils; a high yield per acre; good storage capacity; and high nutritional value. Among the varieties grown in the 1840s were the Apple, Cup and Black but the most popular was the Lumper. By the 1800s it had replaced other foodstuffs and, with buttermilk, became the main component of poor people's diet – the average daily consumption was fourteen pounds for a man and eleven pounds for a woman.

By 1845, Ireland had a population of 8.5 million, of which 4.7 million had potatoes as their main food source. This over-reliance proved devastating when blight struck. This fungal disease, *phytophthora infestans,* managed to destroy a third of the national crop in its first year. The grim death of one million people, and the emigration of well over a million more, resulted in Ireland and the potato being forever linked.

One of several explanations for the famine was that a climate variation brought warm wet weather to Ireland which, in turn, allowed the blight to thrive. Now a change to our weather system threatens potato growing altogether: global warming.

According to the Environmental Protection Agency, the east coast is going to experience much drier summers. This means the tubers will not have the necessary moisture to expand. Put simply, they may not survive. If farmers have access to a local water supply, they may be able to irrigate their land. Those who don't face paying large amounts for water. This would be expensive and make potato-growing uneconomic to pursue. It's a stark prediction when you consider that the top three producing counties come from the region most under threat: Meath, Dublin and Louth respectively. Wexford comes fifth in the league.

Farmers such as David Rogers would be directly in the firing line if climate change projections turn out to be true. Along with his three brothers, he farms more than 2,000 acres of cereals and 400 acres of potatoes. "We start harvesting in July with an early crop of British queens, then Kerr's pink, but the majority is made up of rooster, which is now Ireland's most popular potato." Their early potatoes go to traditional markets in the west of Ireland and are then distributed around smaller shops in the region. The bulk of the potatoes go to a company called Country Crest in Lusk who wash and pack for Tesco Ireland. "The yield will vary from ten to eighteen tonnes an acre. The later crop would be the higher yielding."

The large range in yield is accounted for by weather conditions, particularly during planting season and the final growing period. "If you get a wet spring, the soil will not be as good and you will be late planting. Also, if you get a dry summer, your yields are reduced as there isn't enough moisture." So what, I ask, would be the perfect conditions for growing potatoes? "The best would be a spring where you get a couple of weeks of reasonably dry weather and then get moisture in May and June. You could handle a dryish summer then."

The summer of 2007 was the most peculiar David ever experienced. "From Saint Patrick's Day to June, we didn't get any rain, which was ideal. We never planted in such good conditions. Then the rain came, which was fine. But it didn't stop. The fields got completely water-logged; the roots were sitting in water and just didn't develop as they should. Then it picked up in the middle of August, and we got ten dry weeks again, which was ideal for harvesting. It was the most peculiar year as we got three cycles: ten weeks of dry, ten weeks of wet and then ten weeks of dry again." This weird weather contributed to a lot of blight and a twenty per cent drop in yields.

Yet David didn't see this as necessarily a harbinger of a dark future – he remembers previous exceptional years. "Most people were equating 2007 to 1985 and 1986 – two years with very wet summers. I was equating the dry spring to 1993, the year I got married. It stayed completely dry for around ten weeks as well. I went off on my honeymoon and when I rang back, to see how things were going, I was told the place was flooded. Those trends have been there in the past. Older people remember, I think, 1958 or 1959 when there was severe flooding. Every year throws up something different. It makes this job interesting. We can't farm to dates as such."

Yet David also recognises that temperatures are, in general, on an upward trend and says this is particularly visible in winter. "Winter definitely isn't as cold as it once was. Growing up, we remember the icicles, but you never get that prolonged frost anymore. You seem to only get a frosty night, but the next day turns mild again. Last year, April and May, as well as September and October, were warmer than normal."

If the summer temperatures do increase, David is lucky as he has access to a water source. "We have a river here – the Ballyboughal." With a smile, he quips: "We didn't need it last year." But David and his brothers have examined what they might have to do if trends continue upwards. "Long-term, the first thing we would do is to bore wells. We have a big aquifer here." This water-source extends from Swords to Balbriggan; however, there is a difficulty. "The biggest problem is that Fingal County Council plans to build a dump on it. They are saying they're going to line it and it's going to be safe. But it's a time-bomb. All the experts in the world can say it's safe, but when it leaks . . . where are the experts then? They say they're going to monitor it but if the dump leaks it's ruined."

An added problem for irrigation is the way modern farming is conducted: while the Rogers family farms nearly 2,500 acres of land in north Dublin, they own only twenty per cent and effectively rent the rest. What has made this new system possible has been a lot of younger people leaving the land. "We would have taken over several farms in the past few years. It's very common in north County Dublin."

There are a number of implications from so much land being rented – the key one financial. "Because the price for grain is high, the guys setting the land rental price are looking for more. It also means more lads want to grow grain and so there's less land for potatoes. For cereals, the average rental price would have been €110-€120 an acre last year. It would be €150 or €160 this year."

Introducing systems of irrigation is not too difficult, as David found out while visiting a farm in England. "They had a reservoir. It was about five acres and, during the winter, they pump water into it from the river. If it takes four weeks to fill the reservoir, they know they have four weeks of pumping back out onto their crops in the summer." However, he recognises that this is unlikely to happen on rented land. "The truth of it is that you're generally rotating around on land from different farmers who don't grow potatoes. That will become trickier because you won't be able to get a reservoir on the rental land, as you might not be in it next year."

David's family only got into commercial potato-growing when they started to expand the farm twenty-five years ago. "My father would have grown a small amount of potatoes and sold them locally. At that time, my older brother and I had come back on the farm. It was a way of having more work on the farm and creating more income. It just grew from there." To spread the harvest season, they also grow winter wheat and barley. Growing the potatoes, though, is fairly straightforward. "We wait 'til March, until the weather settles down, and then plough the fields. You go out then and rotavate the ground and form it into big ridges. A machine, called a de-stoner, then goes through that ridge. The stones and clods are pushed into the alley and you're left with a fine bed. It's probably one of the best inventions for potato production because now when you go to harvest from your drills, the clay falls through the harvester and only potatoes go into the trailer. Before, you had to have an army of people picking all that."

Yet despite the successful business which he and his brothers have developed from potatoes, they are re-examining their planting plans. With a fall in demand for potatoes and an increase in price for grain, things are going to have to change. "We're cutting back our acreage this year, because it's just getting a little more difficult. Up to now we were expanding as others were going out of business. I think we've peaked now and will be reducing." I ask him by how much. "I'd say we'll be cutting nearly twenty per cent off for this year."

If the family is reducing its focus on potatoes, I wonder whether the new emerging green market might be an answer. The brothers are "looking into it" but were not particularly impressed when they initially dipped their toes in the water of the biofuel market. "Last year we had 250 acres of oilseed rape, but the return wasn't as good as the wheat. This year we didn't grow rape, but lads who did lost out because it didn't rise in price while the wheat did. But it all comes down to the market and what you think is going to leave you a profit."

This issue of financial return is a "big concern" and a big influence on farmers' attitudes. Given his experience and strong grounding in the Irish Farmers' Association,

David feels the government needs to do more to make the green market appealing. "The government is only barely giving enough incentive to farmers to get into it." It should, in his view, be offering certain guarantees. "If you are going to produce X amount of rape, then contracts should be given out so that you know, before you sow, what you're going to get. People then would have confidence." I ask him if a guaranteed price would ensure a guaranteed market. "I think it would make a huge difference."

He also felt let down when, with rising oil prices, he tried to grow oilseed rape with a view to fuelling his tractors and other farm machinery. "We looked into doing that with another farmer about three years ago. We had it all set up, submitted a plan to the government but were turned down." There's consternation that this is the case when farmers already pay reduced rates for diesel over road-users. "We're already getting the benefit of that. I can't see why the government couldn't allow us grow a small amount of rape, crush it and use it ourselves. But if you get stopped by Customs and Excise officers, who dip your tank and identify that you didn't get the fuel from an official source, you're breaking the law." Does he feel it all comes down to government fears of a drop in tax-take? "Oh, I'd say so. Definitely."

Supporting the biofuels market sounds straightforward. But critics point to the low yield per acre. They also identify a problem with the amount of energy that's required for converting the seed into oil. There's the added complication that growing rape means less land for cereals and higher grain prices. Indeed, the International Monetary Fund estimates that biofuels are responsible for thirty per cent of the price hikes in the past three years. Another concern would be guaranteeing a price for an agricultural product which could collapse, leaving the taxpayers stung.

David quickly points to the fact that the price of fuel in Ireland has "rocketed up" and is causing immense strain. "It's putting a fierce expense on running tractors and, especially, drying corn in the burners." For him, it seems strange that a local option is available but no-one seems to be taking it up. "A lad was telling me he was away on holidays down in the west. He saw a German in a VW pull up outside Lidl. The man came out with two crates of cooking oil and poured them into the fuel tank. When my friend asked him, 'What's going on?', he replied, 'This is the cheapest fuel in Ireland.' The Germans are big into bio."

Yet maybe the International Year of the Potato will turn things around for an industry estimated last year at about €80 million and involving around 600 growers. The IFA is coordinating a national plan of events, with nutrition a central theme. Young people are also being targeted through a special initiative called "Meet the Spuds" in 4,000 schools and through a dedicated website. This is a top priority, given that things have changed so much since the days that a dinner wasn't a dinner without potatoes. Heritage centres across the country are looking at the role of the potato in Irish culture and its benefits today.

Organisations such as the Irish Seed Savers Organisation are also encouraging people to grow more native potatoes from seed. Log onto their website, and one variety available dates back to the famine – the lumper.

In straight-talking prose, the website informs you, "Flesh is white and waxy but taste is poor and culinary use limited." A far more appealing "early crop" is Sharpe's Express from 1900: "Came top in our taste test of the first earlies. It is a long white variety. Yield and disease resistance are poor but flavour is excellent." From the main crops, there's the Arran Victory, which dates to 1918 – a round blue-purple skinned variety with a floury texture: "It has good cooking qualities and good flavour. Growth is vigorous and yield is heavy."

But in this modern era of "time-poor" families, what will probably make the difference for potato-growing in Ireland is the ability of the industry, in the short term, to break into the convenience food sector. With people increasingly purchasing ready-made meals or dining out, it's up to potato growers to ensure their product is part of the new make-up. The Central Statistics Office says that, in 2007, potato imports amounted to 607 tonnes, valued at €537,000. A lot of this would be accounted for by convenience food, like frozen chips.

This drive to satisfy the consumer also has an environmental knock-on. One very good example is the rise of the rooster potato. People want to eat them, which is fine. But, as David found out, the problem is that they want to eat them all year round. "We usually plant the crop in April and harvest in October/November." According to the Department of Agriculture, forty per cent of the land for potatoes is accounted for by roosters. So where do the Irish roosters I buy, say, in July, come from? "They've been in a fridge for six months." In times past, farmers would grow golden wonders as they had a long growing cycle. Not any more. "We have lost seasonality." We also have a big carbon footprint as it takes a lot of fuel to keep fridges running for six months so that consumers can buy roosters in the summer.

However it's not all doom and gloom. David has found that the market also has the capacity to drive environment improvements on the farm. His relationship with Tesco Ireland is a good one. "We're in a scheme called 'Nature's Choice' which brings a whole load of criteria. You're meant to recycle paper and plastics, but also things like batteries. It's no use saying 'a lad came and collected them' – you have to have the cert. Before this, people might not have thought about waste oil and emptied it out the back. Now you have to have a special tank. All the EU payments are becoming more environmental – 'The reason you are getting this is because of these conditions.' It's making farmers think – we need the payment, so we'll do it."

But will he stick with potatoes when projections suggest climate change could eliminate the crop in just a few decades? "If you look at Ireland, we just get so much rain. We're a country of rain. There's an awful lot of places in Europe, like Spain and Italy, that would presumably dry up an awful lot faster. In the short term, this country could be the food basket of north Europe." While he sees the positive, and says climate change doesn't strike "fear into the heart", he accepts, as the decades slip by, that this might change. "Going beyond that, it gets scary. If the production of food becomes a problem in northern Europe, where in the world could you produce it? The more it's in the media, you do begin to think 'wow'. I suppose we are all going to have to wake up to that and adapt to it."

Attitudes, like the climate, appear to be changing. Ireland used to be the largest consumer of potatoes per head of population in the EU but, with the entry of countries from the east, we've been passed out by Latvia (147kg), Poland (126kg) and we now rank third (123kg). But it'll be a long time before the potato is no longer associated in Ireland with things like *poitín*. Despite Germany being the biggest producer of potatoes in the EU, it will also be a long time before Irish people are no longer associated with the potato – whatever about consumer choice or climate change.

A few years ago, my friend and RTÉ colleague Mark Little got an insight into how Irish people are still inextricably linked with the humble potato while he was working in Boston. Mark was reporting on how Irish high-tech companies were securing an economic foothold in the city. Under a bit of time pressure, he found himself arriving into the harbour district late and, with no free spaces, parked the car where he shouldn't have. The cameraman left a note on the dash stating they were reporting on a story for Irish TV. Given the strong associations between the green city and the Emerald Isle, he underlined the word "Irish". Just to be sure. When the interview was over, Mark returned to the car only to find a note had been left under a windscreen wiper. The neatly typed script read: "Who gives a flying fuck who you are . . . you're illegally parked, you potato-sucking moron."

17

Business to
the Rescue?

Declan Murphy knows more than most about the impact
of climate change. He lives with his family on a fjord in
Norway. "My house is surrounded on three sides by
water. During the last high tide, when there was also a
storm, I measured the height of the water. It's now only
two feet from my back door. It's quite alarming."

Declan grew up in Limerick, a city he describes as the
wettest in the British Isles. Now he resides in the city
which records the highest rainfall in Europe – Bergen.
"It's a city built around mountains, so it rains really heavily.
The rainwater charges down the mountainside." The
population of around 350,000 is also beginning to notice
changes. "You very rarely see a flood in Norway because
of the massive drainage system. But twice this winter,
Bergen was brought to a standstill after the entire transport
system was flooded. The volume of rain was so intense
that not even Norway's drains could cope with it."

When I meet Declan Murphy for the first time, I find him a little confusing. The pinstripe trousers signal his business background, but his reddish locks flow down onto the embroidered leaf-patterned collar of a crisp white shirt. There's an unassuming and likeable quality about him, but his background is pretty unique and his message is quickly gaining ground and influence.

He started his career working for a mechanical engineering company which clearly spotted his ability and flair. "Fresh out of college, they gave me a blank piece of paper and said: 'We have a company in the US, go and run it.' I went to California for three weeks and ended up staying three years." The business was selling technology which could tell if fruit was ripe or not. After making the product more robust, Declan changed the marketing. "We started selling direct rather than through agents and built long-term relationships." The trick worked and the company moved from being loss-making to "hugely profitable".

In 1991, he moved out on his own, setting up a software upgrade company in London. In what would be a "baptism of fire", he lost control to his financial backers after two years. Declan returned to Dublin just before the tech boom and, undaunted, got into a banking systems company. Over five years, he helped it move from turning a profit of €12 million to €50 million. "It was about seeing the hidden value in the company and unlocking it."

The big change happened, however, when he ran into a stranger at a party hosted by U2 manager, Paul McGuinness. "We just got chatting and I really liked him.

He realised I was involved in tech stuff and asked me would I look at some small investments. I had a look a couple of days later, saw they were not so small and suggested he do x, y and z. A month later the phone rang and he said: 'I did what you said and it worked. Do you want to do this properly?' I became his private investor."

Declan is reluctant to identify his backer – "names don't matter". However, it was a dream business job: he started to buy up stagnant companies around the globe and then try to realise their hidden value. He admits there were some "spectacular failures", but overall it worked out very lucratively. Yet there was also an ethical core – with every commercial deal, they would also organise a philanthropic venture back-to-back with it. "He was sending ice-breaking ships to the Antarctic, which he gave to Greenpeace for free; he was sending ambulances to Cuba; he was building hospitals and schools in Africa; he was buying land-banks to stop developers destroying natural habitats; lots of really well-picked social projects."

The mysterious investor retired a number of years ago, but his approach has clearly rubbed off on Declan. Their success also meant Declan could effectively retire too. He's now involved in setting up his own social projects. "They are things I'm passionate about and see a need for." One of them is called Forest, based in Wicklow, which helps people with alcohol addiction or drug dependency. "Last year 120 people attended and we got excellent results. I guess the reason I did it was that I helped a friend go through it. I saw what the treatment methods were like and was shocked. I felt we have to be able to do something better." Declan describes it as "a business with a heart" – "It's able to help some people who can't afford it, but also charges those people who can."

There was no revelatory moment at which he can pinpoint when global warming became the single most important issue in his life, apart from his family. "It was just the general media attention. So I started to read about it." But once Declan decides something is of critical importance, things happen. In the case of climate change, this occurred in April 2006. "I just got stuck into the subject. It has really steamrolled since then." His wife says he's found a new religion.

One of the things he did that summer was to become an advisor to the executive of Greenpeace in London. While he retains a huge respect for the organisation, their partnership didn't work out. Breaking point happened when Marks and Spencer started a green advertisement campaign and used the Greenpeace name without permission. Declan contacted the massive retailer, which said it was prepared to give Greenpeace nearly a million pounds. "Marks and Spencer were prepared to write a cheque to say, in part, sorry for not asking; but also to show they supported what Greenpeace did. Yet Greenpeace couldn't take the money because they've a charter which says they can't take corporate donations. In one way I understand that principle, but the time for it has gone. Climate change is happening fast and we don't have time to wait. Frankly, we need the business community to get involved."

This is Declan Murphy's main credo: business and industry must play a fundamental role if we are going to find quick but lasting solutions to climate change. "I found that some of the campaigning organisations were just not able to engage with business people. There's often an ideological blockage. Yet the business world is crucial to the delivery of solutions. When you look at our society, everything is related to a business somewhere – the clothes we're wearing; the coffee we're drinking; the table we're sitting at; everything in our world touches off a business somewhere. So we'd better get the business community onboard. Capital is the only thing that can get us out of this mess."

While he was disappointed that his relationship with Greenpeace hadn't worked out, the experience gave him the idea for the role he could play: getting business into the centre of the debate on climate change. The first step was to read up on the hard data. And the more he read about global warming, the more concerned he became. It wasn't just the scale of the problem which was ringing alarm bells, but also the pace of the political response. "What drives my timetable is the physical scientists and what they are telling us about our planet. Frankly, all other timetables are irrelevant. When I speak to the physical scientists from the UN's climate change advisory committee, the IPCC, they say it's highly unlikely we can avoid the planet heating up by more than two degrees Celsius, and irreversible impacts will follow. The most important thing, therefore, is to continually bring that physical science timetable to all discussions so that our decisions reflect that reality. Otherwise we are goosed."

This is where Declan's business outlook comes into something of a conflict with the political world. It's clear, he says, that once you have a timetable for action, there should be a detailed step-by-step plan identifying how you will get there. Given that the future of the planet is on the line, he finds it nearly impossible to comprehend why governments don't do this. "For our successful entrepreneurs, this is precisely what they do all the time. If you give them an insurmountable challenge, they'll get so stuck into it. They will focus on results. They will mobilise resources that have not been brought to bear on an issue. We need an injection of creativity to find the solutions and can-do attitude."

Declan's next step was to set up an organisation to bridge the gap between the worlds of business and the politicians. It's called The Ecology Foundation. His first meetings with government departments, like Environment and Energy, have been very positive to date. The officials certainly like him. Maybe talking business makes a pleasant change from being "harangued", as they will put it, by campaigners.

But Declan has very clear ideas about what, how and when things need to happen. "You need to introduce business people to the debate who are used to facing up to an identified challenge. When I ask politicians to show me their project plan, they look at me in sort of disbelief. In the business world, that's the first thing you would do so you know precisely what elements you have to deliver. We're missing that at the moment."

One of the first steps for the Foundation has been to try to get into the boardrooms of Irish companies and to speak their language about how global warming can affect their business and what opportunities are coming.

"What businesses need is someone to give them a very simple framework and maybe a little bit of a push to get started. I've been into a hundred boardrooms in the last five months. Of that, only two companies had their heads in the sand. For the most part, I found the chief executives are very aware of the issue and know they are going to have to do something, but have not had that last push."

Possibly one of the reasons that Declan Murphy gets easy access to these boardrooms is that his message is not one of doom and gloom. Instead, his language is peppered with words like "positive", "possibilities" and "potential". "This isn't an expensive process for companies to engage in. It's very low cost, but there are massive benefits, which are not simply limited to energy savings. There's also staff motivation and the retention of customers. If you don't adjust to this new world we're entering, you will be left behind."

He is pushing for 500 small and medium-sized companies in Ireland to join a club which has an overall aim of reducing their CO_2 emissions by 50,000 tonnes within five years. That's equivalent to taking 8,000 average-sized cars off the road. "The reason for forming a club is to prevent companies from feeling their actions are not important – they are. It also provides an opportunity for their efforts to be recognised." When I put it to him that 500 small and medium-sized companies can only make a small difference, he shoots back: "But if you then had the Fashion 500 and the Construction 500, you might get places." Declan then illustrates how a company can exert its influence in a very powerful way. "The US retail giant Walmart has imposed an energy specification on all its suppliers. The spec must be complied with within two and a half years or they will no longer be suppliers to the company. The impact from that is absolutely colossal. Even for small companies, you can achieve the same ripple effect."

Another initiative getting off the ground is called Climate Campus. Currently it consists of a research group which, it is hoped, will expand from ten to thirty people. The plan is also to have a business school: "We will train managers to be climate and sustainability aware. The aim would be to turn out graduates with Masters in Business Administration. Hopefully they would have the acumen to know where the commercial opportunities are going to lie in the next twenty to thirty years. There's a real shortage of climate-focused business people at the moment, even though that's where so much of the world's investment funds are going." If the graduates do have good ideas, a third component would be the opportunity to secure seed funding.

To most people, climate change is something that will only have consequences many years down the road. But what Declan has been pointing out to companies is that they are making decisions today that could have catastrophic financial impacts for them. "You may be about to buy into a company which has a high dependency on transport. Before any agreement, you have to examine the climate consequences. For example, if a carbon tax is introduced in the future, it could make the business financially unviable. You had better make sure you cover your rear before paying out millions for a company. You could be buying yourself into massive risks, similar to the problems with asbestos thirty years ago."

The Ecology Foundation argues that the government must put incentives in place so that business can have an impact. Given Declan's background, it's to be expected that he sees massive opportunities when everyone else is talking about problems. The current fall in house prices and a slowdown in housing completions is a good example. "I'm amazed when the construction industry talks about a downturn in building and cries doom and gloom. They should be out jumping up and down about climate change – it's their salvation."

Declan's big idea is simple: it's predicted that climate change will result in massive flooding around the country, so why not put the infrastructure in place now to deal with the future problem? "It's just one example of the many large-scale infrastructural developments we have to do in Ireland. To upgrade our drainage and river defences, and without even taking our coastal defences into account, is probably close to the scale of our motorway-building programme. That's going to give the government a great capital investment headache, but provide the construction industry with a huge boost." To Declan, it's just logical. We know increased flooding is coming, which will be costly. The Office of Public Works already has extensive plans for the areas needing attention. The insurance industry is begging for action to be taken. Why not simply do it now rather than risking greater expense later?

Declan does not limit himself to striving for change in Ireland; he's also trying to see how we could punch above our weight on the international scene. He has an idea for a green version of the International Financial Services Centre in Dublin's Docklands. It's estimated the IFSC administers around €1.3 trillion in global investments. Part of its secret of success is that it offers a rapid regulatory approval with very high standards. "What's happening around the world is that more and more investment flows are trying to get involved in climate-related projects, such as alternative energies. We believe we could create a green sub-set of the IFSC, which provides similar incentives to attract these new funds. They already have all of the other service resources – lawyers and tax advisors – we just have not packaged it."

What's eating at Declan Murphy and his foundation is that Abu Dhabi has already begun to put such a system in place. "Why not us?" he asks, when this green financial market, sometimes called "the carbon market", is about to take off? "Some people are predicting that carbon trades will exceed cash trades in the next thirty years. The market for carbon trades is in the trillions. We have all of the necessary components here in Ireland right now. It would cost us €20 million to package it and we would be projecting an image to the world of innovation in this new green space."

It does not end there. The Ecology Foundation is calling for the business sector to be allowed take a greater role in international negotiations. Declan believes if business leaders were given a role in the talks processes they would ensure real results were delivered real soon. "What I would like to do is to introduce some more business problem-solving approaches. There's a target to be hit and nothing is going to stop us in getting there. The planet isn't going to respect our deliberations or our need to build consensus. We don't have the luxury of time. We should assemble a team of entrepreneurs from around the world to attend the next UN environment conference."

I wonder what type of individual might give the often excruciating talks process a kick in the backside. Declan laughs. "I'd bring Denis O'Brien. I think he really would take no prisoners in the negotiations. I don't think that he'd listen to any nonsense about wording. We have a number of very successful business people who have a proven record in cutting to the chase. It's the end-game that's really important."

He also has firm views on how to unblock the log-jam over who should pay for new technologies needed by developing countries to reduce global emissions. "It's not beyond the wit of man to create a super-fund of trillions, which is managed by the World Bank or the IMF, and pays for clean energy solutions in countries that so desperately need them. If we don't short-circuit this 'who will pay' question, we'll simply be losing valuable time. Every week in China, there's two more coal-fired power plants. Those plants should be fitted with the absolute cleanest technology available today. It is available today. If we had a fund, we could pay for it today."

Talking about trillions, super-funds and global deals comes easily to Declan Murphy. To the ordinary person, the solutions can sometimes seem nearly as scary as the problem. In his view, though, one of the key weapons of a successful entrepreneur is the freedom from fear. "In the business world, people are used to being scared of risk, but still having to deal with it. You are always scared of your competitors. You are always scared of lots of things you are juggling. Yet you have to have an ability to compartmentalise them, deal with these fears and still make clear-cut decisions. I don't think the business world is switching off climate change because they are scared of it. I think it's just looking for a little bit of clarity on how it can engage."

I am intrigued to find out what Declan thinks about the performance of the current coalition government. In late 2007, Brian Cowen, as the then Finance Minister, surprised many, and certainly me, when he announced that the environment was going to be one of his top three priorities while in government. The then Taoiseach, Bertie Ahern, had previously announced a "green revolution". There are relatively new government policies on reducing our emissions and switching to more renewable electricity generation. What does Declan make of these? "I don't think the documents take account of the science. There's a timetable in which those decisions have to be taken. Otherwise we will simply miss the boat."

Declan has made what some would view as a jump from capitalist to environmentalist but he insists the two can co-exist. It's eminently possible to save the world *and* make money doing it. Some statistics bear this out. Part of the green market, which is defined as the Environmental Goods and Services Sector, is growing rapidly each year. The global EGS sector was estimated to be worth €745 billion in 2004. Research by the British government predicts that the global market for the EGS sector will increase by between a third and a half over the coming decade.

Yet it's the issue of time rather than money which is a constantly recurring theme with Declan Murphy. "I don't think people understand just how little time we've left.

Urgency is vital. It's not that the world is going to fall apart, but our ability to correct the impact of climate change really lessens if we go beyond fifteen years." At the same time, he is optimistic. "You can get depressed about it, and I do on occasions. Yet it's also a huge opportunity for people. They should look with an entrepreneurial eye at where there might be a space in the gold rush that's going to happen."

18

Falling into the Sea

"You can see how the sea has eaten through the cliff and come out the other side of the headland." Gerard Whooley has to shout to compete with a gust of wind at Blackhall in south Wexford. As he gestures for me to follow him to a ledge, to see a good example of coastal erosion, he reassures me: "It's quite safe." His friend, Karin Dubsky, is nonplussed and opts to stay by the fence. "Walk carefully behind him," she advises.

Gerard strides confidently to the edge. "If you just step out farther, onto the point here." Gingerly, I do as he says. But, as I reach him, Gerard looks puzzled. "You can't actually see it now. The sea must have eaten farther back."

There's one thought in my head: if the sea has eaten away the cliff face beneath us, then maybe where we're standing isn't the smartest place to be.

Gerard is still trying to figure out what had happened. "I used to be able to come out here and look through the sea-arch. It's only a distance of about twenty feet, but I can't get out far enough."

I suggest hopefully: "Maybe we'll trek back so?"

As we head back towards the path, Gerard is now bemused. "I've not been down here in three or four years and it's already changed." Coastal erosion, it's clear, is not a theory but a reality.

Gerard Whooley was born at Lough Duncormick close to Bannow Bay in south Wexford. More than three decades ago, he joined the Irish Coast Life Saving Apparatus, known nowadays – after four name changes – as the Irish Coast Guard Service. Now sixty-seven and retired, he still takes a keen interest in what's happening in his area. With Karin Dubsky, another Wexford resident and also the head of the environmental pressure group Coastwatch, we are taking a tour of the coastline to look at evidence of the power of the sea and its association with global warming. Blackhall is a good spot, nearly halfway between where our journey began at Bannow island and where it is due to end at Cullenstown.

A few minutes before our cliff walk, we had been standing on an old road near Blackhall which snakes its way around the coast. Well, it used to snake its way around the coast. Now, a large section has collapsed as a result of the sea crashing relentlessly below. We stand at a barrier and look across to where, about seventy feet in the distance, it is possible to see the road again. Between us and that point, it appears as if a great triangular bite has been taken out of the cliff. Gerard knows this place well. "You can see that the barrier, which was put up by the County Council on the other side, has also fallen into the sea." Karen chips in: "And the sign to tell you about the road collapse has also gone." Part of the footpath had also slid down into the churning salty water. Gerard points to the broken road. "Within the past ten years, maybe even less, I would have been driving along here."

Gerard calls this a "sorry place" where nine men in lifeboats from Fethard died trying to save the crew of the schooner *Mexico*, off the Keeragh islands, nearly ninety years ago. "You get fierce storms along here in the winter." What's worrying Karin Dubsky, who works at Trinity College Dublin, is that the storms are occurring "with a greater frequency now". I wonder if this is just anecdotal but she has evidence in the form of hard data from Met Éireann. "This isn't my imagination. This is what's happening according to people who measure it. We also seem to be getting longer dry spells and longer wet spells."

Both Gerard and Karin accept that the rate of loss is particularly high on the south-east coast. Karin puts this down to many challenges coming together. "Much of the coast is exposed to the open ocean and the waves can get a run at it. Additionally, the land is still sinking since the last Ice Age, adding to the sea level rise which is being tracked across the world. As if that wasn't bad enough, much of the south-east is soft sand, boulder clay and soft rock which offers little resistance to an angry sea. Spectacular losses can be witnessed when heavy rainfall teams up with high spring tides. Chunks the size of houses can cascade into the water."

To the uninitiated, talk of coastal erosion being a pressing problem might seem over-the-top. However, according to the Department of Agriculture and Fisheries,

1,500 kilometres of the Irish coastline are at risk from erosion – that's twenty-five per cent. The counties most acutely at risk are Galway and Mayo in the west and Wexford and Wicklow in the east. It's estimated that 975 kilometres are in particular danger. The department is currently in the process of estimating just how much land will be lost by 2050 and hopes to have the statistics ready by the end of 2008. A series of government-commissioned reports found that parts of the east coast can lose up to two metres of land a year.

And that could get worse, according to Dr Gerard Farrell, Chief Engineer at the Department. He recently stated in a research paper: "In the climate change scenarios . . . mean sea level is predicted to rise and the frequency and severity of coastal storms is predicted to increase. These consequences of climate change will significantly increase the risks posed by coastal erosion and coastal flooding."

My journey with Gerard and Karin began at Bannow island, an internationally recognised world wetland site and a wonderful migratory bird habitat, with salt marsh and mud-flats. It's a remote, beautiful spot which is right beside the reputed chief landing place of Strongbow's Norman invasion force in 1170. As well as being historic, it's also a momentous place in terms of sediment movement – there used to be a town of Bannow, but it's now under water and sand. The town returned representatives to the Irish Parliament until the Act of Union in 1801 but, Gerard tells me, the only structure that remains is the church. "It's peculiar for a church as it has battlements."

Sited in an area with copper and silver deposits as well as an abundance of shellfish, it was also located in a place where the sea could wreak havoc. It's a story which fascinates Karin. "There are records of the last time the MP was elected. He was holding the chimney pots of one of the town's houses, but everything else was already covered."

Today Bannow island isn't an island – it's connected to the mainland by a narrow sliver of land, known locally as "the neck". However, it is vulnerable to sea level rise and exposed to storms. It's quite possible that Bannow island will become a real island again. Gerard's house is just five miles away and the changes in the area have been immense. "I can remember as a lad, you could walk for hundreds of yards on flat smooth sand. Now it's a cliff face. It's not level anymore but on a strong incline."

As luck would have it, Karin says this highly endangered area also happens to be home to one of the country's most endangered plants. "This is a very rare species, a perennial glasswort found only in Bannow Bay. Here on the neck, the plant is very much in danger." Since this has been highlighted, many local people have become interested in protecting it, even those, like Gerard, who had not been aware of its existence. "Prior to meeting Karin I wouldn't have. I could have been standing on it and wouldn't have known."

The plant itself is a survivor. Fourteen years ago, Eugene Wallace conducted a survey of the plant at six sites and found that ninety-seven per cent were growing on mudflats. As these areas have shrunk and disappeared, Karin says the plant has managed to move. "In all cases, the plants have now left their former sites and are living higher up in the salt marsh. Other plants are just drowning." The perennial glasswort moves by sending out a branch towards the area it wants to go, and then roots like a strawberry. "All its new rooting is away from the water's edge."

One of the problems facing the plant is man-made. Karin notes: "We have already lost one of the sites due to quad and car racing. We have pictures of it when it was there. Grass comes back, but the rare perennial glasswort will not."

The motion of cars on the dunes also makes it easier for the sea to erode the sand dunes. The tyre tracks create perfect channels for the salty water to take hold. Gerard explains: "If the sea takes the dunes out of here, well, that's nature taking its course. There isn't much you can do. But if it's destroyed by man, or by the actions of man – it would be a crying shame."

Gerard Whooley knows a lot about the power of the sea. "I spent thirty-two years in the Coast Guard service." This was a land-based job in which he sometimes patrolled beaches. On many occasions, he was called out when a ship got into trouble. He would bring what was called the "breaches buoy" – effectively a type of harpoon which would be fired from a beach into the side of a vessel. "You had a rocket about three feet in length, with a nine-foot-long wooden stick and length of hoser rope. The rocket was quite a powerful thing and could go out as far as three-quarters of a mile." Once a line was attached to the stricken vessel, the crew could be hauled ashore one at a time.

Such experiences have made him view the power of the sea as "very dangerous". Since his retirement, he continues to watch as it erodes places he knows well. "In Cullenstown, we have a handball court. It's now by the water's edge but, back years ago, was well away from the sea. Well, one night the sea knocked the wall down. I was actually there when it happened. One wave just took the whole wall down and pushed it twelve feet across the court. There's fierce power in it."

It's one of the reasons that he got involved with Coastwatch – now he watches out for the environmental group rather than the Coast Guard. He attended his first public meeting after hearing Karin on South-East Radio, appealing for volunteers. "I just came along, being nosey I guess. I thought I would see what the group was all about. I got into it then." What, I wonder, do his friends think of him now spending his time trying to protect plants down at "the neck"? "Most people I know would be quite interested in it. They don't want to see it lost – especially being so rare and from this area. I don't know too much about flowers. I didn't even give the wife a bunch for *that day*. But certainly I wouldn't like to see the plant lost. It should be maintained at all costs."

From Bannow island to Blackhall and on to Cullenstown, the sea is like a marauder seeking out the weak spots. In Karin's words, the sea "bounces along the cliffs" until it identifies a softer rock. In essence, the hard rock-face will deflect the force of the wave onto the softer zone, increasing the erosion. According to Gerard, the whole area is a place of unique contrasts. "Where there is erosion and a net loss of land, you may also have deposition right next door. The rocks from Bannow down to Cullenstown are also supposed to be the oldest rocks in Europe, especially around the Saltee islands. At the same time, we have the newest cliff faces."

As we drove towards Cullenstown, weaving away from the coast, Karin outlined what she believes should be the key priority for the government: introducing a fair transparent coastal erosion and wider coastal zone management policy. How would she describe things as they currently stand? "It's an absolute disgrace. The government has given no leadership and local authorities are given no guidance." For Karin Dubsky, it's all highly frustrating. What she calls "unsuitable" hard sea defences are impacting negatively on birds, like sand martins, which nest in sand cliffs; a consultant's report for the government on coastal zone management has been "decorating shelves" since 1997; and more infrastructure and buildings have been constructed along eroding coastlines. She says that in a recent EU review, Ireland came out weakest on coastal zone management. Officials from the department would point to multi-million euro investment, ongoing efforts to establish the scale of the problem and intentions to devise a policy framework to respond to it. Yet it means little to Karin. "If you asked me twenty-five years ago how things stood, I would have said there are great plans. A quarter of a century later, the same plans are there. Nobody wants to bite the bullet."

Karin accepts that most coastal erosion is simply going to happen anyhow. "It would be way too expensive to protect it all and so we should let it happen. But we should also have a national erosion policy so that, in selected areas, you would let the sea come into one area, in order to take the pressure off another." The next priority is to introduce what are called "setback lines" to ensure that houses or other buildings are not constructed close to vulnerable shore lines. "At the moment, we have more and more houses being built close to the shore. We will not have the money to protect them. But when erosion impacts, there will be an outcry as people will say, 'Why the devil did you give us planning permission here?'"

So can we learn any lessons from the county which has arguably been worst affected? Well, Wexford has 264 kilometres of coastline and the County Council's analysis is that about thirty-eight per cent is susceptible to coastal erosion. It estimates that some two square kilometres of land has been lost over the last fifty years. Worryingly, the rate of erosion in some local areas appears to be increasing in recent years.

Dealing with erosion costs money. The government approved more than €1 million for coastal protection works in County Wexford in 2007, of which the Local Authority had to pay a quarter. In 2008, the bill for similar works is expected to be €1.1 million. Flood protection works have also been installed at numerous locations in the county, principally to counteract coastal flooding and erosion. A major scheme is currently being put in place at New Ross with further plans for Wexford town and Enniscorthy.

Wexford County Council deserves credit for being forward-thinking. It commissioned the first Local Authority Coastal Zone Management Plan in the country, in the 1980s. Last year, it appointed consulting engineers to develop a coastal protection strategy and the report is expected later this year. The aim is to allow the Council to plan ahead, rather than continuing with an ad-hoc response when damage has already been done.

Crucially, the County Development Plan includes setback lines of one hundred metres so that construction will not be allowed in vulnerable areas. The distance used to be fifty metres, but was increased in the last plan.

However, the actions of Wexford have not been replicated elsewhere. In Karin's view, the government's failure to have a comprehensive policy on coastal erosion means that small problems are inevitably going to become big problems. "On the continent you have people allocated a part of the shore to monitor. They look for early signs of damage which can be easily and quickly repaired, instead of waiting for the wall to cave in."

Our car rolls slowly into Cullenstown, and we make our way along the coast road. Within a short distance we arrive at Cliff Cottage, a house with one of the most extraordinary façades in the country. Gerard explains the story. "The man who lived there was Kevin French. He decorated the outside of the cottage with sea-shells, in a beautiful patterned design. He picked them all locally." Nowadays, the main road is all that separates the cottage from the cliff. "Kevin had a small boat he used to go fishing in, going back thirty years. He had to push it down on rollers, first over thirty yards of grass and then a further twenty yards of sand before he reached the water."

At one point, as the sea made major advances, some council workers turned up to conduct some protection work at Cullenstown. They were impressed by the house and struck up a conversation with Kevin (who has since died). They asked him if he had found all the shells on the local beach. He said he had and then added: "If you don't get the protection work finished, they'll all be back on the beach again."

We get out of the car to take a closer look. Karin remarks: "It's amazing – it's a mirror of the shellfish bio-diversity of the area, stuck on a house. You've got oysters, scallops, and three different species of cockle. There are also razor shells and two of clams you don't usually see. I've also spotted some whelks, different limpets and a range of winkles. Then beautiful dark-blue mussel shells. It's absolutely gorgeous."

Karin learned to love nature at a young age. "My dad knew all of the bird songs. Every Sunday, our family would go for a walk in the countryside." When she was a little older, she would ride ponies – not to race them, but to get closer to nature. "If you lie down on a pony, you no longer look human to wild birds. We would go right up to Brent geese as they moved on the intertidal zone. The challenge was to see if we could get close enough to see their nostrils. We'd then sketch them when we got home."

While Karin's youth was tranquil, her parents' lives prior to arriving in Ireland were anything but. Both were Second World War refugees – her mother from Riga and her father from what was western Prussia. Karin and her brother were born near Bonn. Around 1964, the West German government offered a compensation package which included a provision that, if refugees were prepared to emigrate, the amount would increase. "My parents were not too happy in Germany, so they took the package." The Falkenthals ended up in south Wexford. Later Karin married Paul Dubsky from north Wexford. One of their four children, Eoin, is now involved in combating climate change with Greenpeace International.

Given her interest in nature, it's no surprise that Karin made a career out of it. She eventually studied for a PhD in environmental science at Trinity College Dublin. To supplement her income, she was also lecturing and making ice cream. What turned her into a campaigner, though, was the formation of the Dublin Bay Environment Group, which was concerned with sewage pollution. The big stumbling-block was getting access to data from the Council. Through sheer persistence, they eventually got the data and found it was starkly out of kilter with EU norms. The whole thing snowballed – including making a programme with RTÉ's *Today Tonight* and getting involved in deciding what sort of waste management system should be put in place. Despite what she described as a "sewage holiday" or fact-finding mission to Holland, her preferred choice of a sewage treatment system for Dublin Bay wasn't chosen by the Council. Given the immense problems that have afflicted the facility since then, maybe they should have listened to her.

The Coastwatch organisation was born out of a dinner with *Irish Times* Environment Editor, Frank McDonald. They decided on a "fun challenge" for the European Year of the Environment, in 1987: to draft a questionnaire on what was happening on Irish coasts and have it published by the newspaper. "We thought we would get a hundred answers but we actually received more than a thousand." The idea was noticed by the European Commission, and Karin was asked to replicate the survey and analysis across the EU. It was to become the biggest coastal survey in the world. "We had well over 10,000 sites in a given autumn and ten times as many surveyors." Coastwatch Europe is still operating today but, in Ireland, much of the time is spent "fire fighting". "So many things are happening and often we're only called very late in the day. We're rushing about and never having enough time to do things like publish." As with all voluntary bodies, money is tight.

Yet the organisation continues to function and is adept at highlighting where the coastal environment and wetlands are being put at risk. Its eyes and ears are Coastwatchers like Gerard Whooley, who had left school at fourteen and, shortly afterwards, ended up in England. "I did everything. I worked in a factory. I worked on a farm. I was at sea with the merchant navy and trawlers. I even tramped the roads for a while." He came back to Ireland in 1973, after marrying Dilys, "a lovely Welsh girl", and has four children. With grandchildren he's "back in the thick of it again", but likes nothing more than to be out, watching. "I just generally keep an eye on the coastline. I'd be having a look around and watching out."

As we prepare to go our separate ways from Cullenstown, I ask Karin what, in her opinion, it will take before coastal erosion is given the priority she feels is needed, given that all predictions of climate change suggest that storms will become more frequent and intense and coastal erosion will increase dramatically. "There will be one big storm, with a big loss of property and life. Major, major property losses. Insurance companies will refuse to insure people, even more than they do now. A disaster? Hopefully not, but, sadly, it's the more likely scenario."

19

New Religion

Sean McDonagh's life was changed, in an instant, by a view. Several decades on, he still gets animated talking about it. The moment occurred in 1980 when he arrived at the home of the T'boli people, who live in the tropical forests of Mindanao, southern Philippines. "I still remember the night. It was an extraordinary place. One of the most beautiful in the world – at 2,300 feet and six degrees north of the equator. And it has a fantastic climate. The T'boli live among some of the most diverse forests on the planet and have three beautiful lakes on one side."

But there was trouble in paradise. In contrast to the beauty of the landscape was the horror of what was happening to it. The loggers had arrived. "As far as the eye could take you – devastation. The forest had been cut, burned and raped." Clicking his fingers, Sean says: "The realisation came to me like that – if we don't protect the natural world, there's no future for anyone." From that moment, he became an environmental campaigner and would never again be afraid – even of his own church.

Sean McDonagh is a Columban priest and an internationally renowned theologian. One of his recurring themes is this: why does the Catholic Church have such a well-thought-out message on social justice and human rights, but has failed to engage with ecology? His forthright view is that the church should be doing much more. He's also highly critical that governments and wider society have failed abjectly to deal with global warming. To him, it's a nonsense to suggest the problem is now mainstream and being tackled. What makes Sean difficult to dismiss is his academic ability to marshal the facts and his deeply held convictions, which were forged through working with "the poorest of the poor" – the people who will be hurt hardest by climate change.

The start to his missionary life in the Philippines, in 1969, was significantly better than that of his predecessors. "We were given a year to learn the language. Ten years before that, someone gave you a grammar book, which was generally full of mistakes, and you would be out in a parish in two weeks." Because of the opportunity, he got to learn the language and culture, which meant he was much better equipped to forge relationships with the people to whom he ministered. That ministry was far more than matters of heaven and hell. "I was very much involved in the fight for justice and human rights. What was great was that, rather than individual priests taking action, the diocese hired a lawyer to look at rights to land. It also set up credit unions."

The political backdrop to this activism was martial law, declared by Ferdinand Marcos in 1972. It's a regime remembered for its violent and brutal suppression of dissent. "One of my friends lost her life. Her name was Maria Elena Ortegas and she was using the Scriptures to empower women to stand up for their rights against landlords and the military. She developed a lot of presentations based on the Magnificat in Luke's gospel. Putting down the mighty from their thrones was not something the Marcos regime wanted to hear so they called her, and many others like her, communists. As such they were disposable. She was killed in 1973."

While ecology and the role of the church would come to dominate his life, the initial focus for Sean was the T'boli. He first worked with the tribe, known as "the people of the lakes", after the local bishop asked him to prepare a plan on how the diocese could help the people. What struck him immediately was what a profound affinity they had with their surroundings. "They find their personal, social and cultural identity within the natural world, which we never do." The forest was part of their lives in a way that westerners can't fully comprehend. "All their poetry came from the natural world. It was the source of myth, food, building materials and religious inspiration. All their songs were about the natural beauty surrounding them. For me, that was extraordinary."

As a child, Sean had a strong feel for the natural world. Growing up in Tipperary, with a teacher for a mother and a garda sergeant for a father, he often fished and worked on a farm every summer. But even that didn't come close to the T'boli's relationship with nature. "They lived it. I could experience it, but not in the way that they did."

Even though the priest and his parishioners worked together, his education and global awareness often set him apart from the T'boli. On one occasion, a wild eagle, whose talons had got caught on fish pens on the lake, was captured by the locals. Thinking back on it, Sean remembers that such a prize was a good meal for an extended family. "Instead of killing the bird, they brought it over to me and we built an aviary. For the next week, the people came to look at this extraordinary creature – three feet tall, with a wingspan of six-and-a-half-feet, with this eagle eye looking through you. What really touched me deeply was the way the people were overwhelmed by the sight. Yet I was the only person there who knew that we would be the last generation of people ever to see this bird in the wild. The habitat which supports it is simply gone."

While his education and worldliness provided insight, they were also privileges and recognised by the T'boli to be such. On one day, he was feeling depressed over the deforestation when the tribal leader met him for breakfast. "It was my second year there, and I remember feeling really overwhelmed by the awfulness of it all. When you went for a walk there were forest fires on the hills." The leader understood something was wrong and so enquired. "I told him I was depressed. He'd never heard of this and asked me to explain it. When I did, he just looked back at me and said: 'That's a luxury.' I asked him what he meant. He said: 'I know that you know where your food for next week is coming from. I don't.' That took me aback quite a bit."

As a missionary priest, Sean was interested in linking his emerging interest in ecological matters with his faith. While there appeared to be very little he could draw on from his reading, he struck gold when he visited relatives in New York and met Thomas Berry, a Catholic priest who was lecturing at Fordham, the Jesuit University. "We had a fabulous discussion for about three hours." He realised Berry was the person who could make the connections between his faith and campaign. "I said, 'I'd like to stay with you and learn – would a week do?' He said, 'No – four months.' That's what gave me the whole framework. What he calls 'the new story'. I ended up taking about half of his books back to the Philippines." Berry effectively illuminated the path Sean had been seeking. "He's the most erudite man I ever met. He's still alive – ninety-three and living in North Carolina."

With little written about the relationship between theology and the vital importance of ecology, Sean committed himself to communicating the message. However, efforts to get published in religious magazines weren't easy. "The title of one piece I wrote was 'The Pain of the Earth' or something. I didn't get a response and was getting kind of angry so I contacted the editor. He said, somewhat facetiously: 'I think it would be more suitable for the *Farmers' Journal*'!" It was a massive struggle to get anyone interested in a book. Sean shows me an album containing twenty-eight letters of refusal from publishers. "It took me three years to actually get a book published. They said they would have to conduct market research to see if anyone would buy it."

In 1986, his first book was finally published in Britain. Entitled *To Care for the Earth: A Call for a New*

Theology, it received critical acclaim and the pioneering work is still quoted today. It was published in the United States the following year. A series of books followed, including *The Greening of the Church*; *Passion for the Earth*; *Dying for Water*; *The Death of Life*; and, most recently, *Climate Change: The Challenge to All of Us*.

Given this background, the first question I ask Sean McDonagh is to evaluate the position of the Catholic Church in Ireland with regard to environmental protection. "In general, there's little thought given to it among the clergy." Yet it isn't all negative. "Religious women are much better. For example, take the Dominicans in Wicklow – instead of selling seventy acres of prime land, they decided to build a place devoted to ecological education and organic farming. It's an extraordinary act of faith in our future. They probably would have got tens of thousands of euro per acre for so-called development. I call it plunderment." Another ray of hope, he says, is how the subject is being dealt with in junior and secondary school education. "Care for the planet is being factored, in a major way, into what it means to be a Christian today, in terms of God, your neighbour and, now, the natural world."

However, his over-riding sense is one of frustration with the actions of the church. "I'm particularly critical of religious leaders. The whole basis of religion is to connect. We should be making the connections." He finds it very difficult to understand why this is not the case, given the scale of the climate change problem. "If we as a race are going to continue, we have to live in eco-systems that are vibrant, and which are sustainable. We tend to forget that and concentrate solely on the human world." The current situation means there is no theological education on ecological matters for religious. "There's not a single educational institute of any of the Christian religions – Catholic, Church of Ireland, Presbyterian – which teaches a single course of ecological theology. You might get something with the Jesuits about ecology and the Bible, but it seems to be something to do if you have nothing serious to think about."

Because of that gulf between what's being taught and where Sean believes the church should be, it isn't surprising that he has often taken flak over his views. He would regularly be asked in the 1980s: "Do you still love trees more than people, Sean?" One response to such jibes has been to write articles about the numerous gospel references to the environment. He has written many such articles; here's one extract: "Many of his parables are centred on nature: He speaks of sowing seed . . . of vines . . . the lost sheep . . . or the life and work of shepherds. His teaching is regularly interspersed with references to the lilies of the field . . . the birds of the air . . . and the lair of foxes. He was Lord of creation and could calm the waves . . . or walk on the water . . . or, when food was needed, multiply the loaves." Later in the same piece, McDonagh quotes Saint Paul to illustrate, as he says, that "The ministry of Jesus was not confined to teaching, healing and reconciling humans and all creation with God. His life and ministry had a cosmic dimension. Paul tells us that he is the centre of all creation: 'He is the image of the invisible God, the first-born of all creation; for in him all things were created, in heaven and on earth, visible and invisible, whether thrones or dominions or principalities or authorities – all things were created through him and for him. He is before all things, and in him all things hold together.' (Col. 1:15-8)."

Inasmuch as he attempts to "convert" his fellow religious to take up the issue of climate change, Sean is also a combatant in the political sphere, both internationally and locally. A good example was when he attended a public meeting in Meath two years ago, at which plans were outlined by Ministers Noel Dempsey and Martin Cullen to bring the railway line to Navan . . . in 2015. He was infuriated that the road-building continued apace while the railway was delayed. "Towards the end of the meeting I quoted a song which goes, 'It's too late to get ready for the past.' I said, 'What you have done tonight is possibly a good response to travel realities of the twentieth century. The two singular issues which will affect travel in the twenty-first century are peak oil and climate change. I don't see the slightest focus on either of those in what you've put up.'"

In Sean's view, all too often lip service is paid to the issue of climate change. "I tend to believe and argue that the subject is not at all mainstream. Did it turn up in the election debate between Bertie and Enda? No." Being direct, he often puts it up to people who contend they are concerned. "I say, 'We're going to have a lecture now on climate change. Here's a piece of paper. Write down everything you know and the gases involved.' You get all sorts of responses." His interaction with the public, and his depressing conclusion, is backed up by one British newspaper poll which found that sixty-five per cent of those questioned don't really believe the scientists that global warming is happening. Why? "I think there are two reasons – the problem of global warming is so big and, secondly, people can't see what they can do about it."

The main problem, he believes, is that it has not yet been accepted that climate change impacts on every facet of life and, accordingly, the response has to be all-embracing. Instead, the life-systems of the planet are being closed down as a result of human activity and no-one is paying attention. It's a theme he takes with him when visiting schools and universities. "If I go into a school, I'll often be told about their green flag. I will say: 'What I'd like to see is your senior economics programme. Where is the economics of climate change?' You'll be told it's in the geography or religion class. I say: 'Are you saying climate change isn't a major economic issue? Have you read the report of Sir Nicholas Stern?' They are effectively teaching the old paradigm – fundamentally, to go out and continue the plunder." What's happening in the schools is a reflection of the political response. "There is no sense of the physical energy which is required to turn this around. Are you telling me they are really focused on this?"

There is a similar situation in the Philippines. Recently Sean was asked by one bishop to plan a religious retreat which was to take place in the capital, Manila. However, he was told the subject of climate change was a "no-go". In the event, he called on the assistance of a friend, a nun. "She has a brother who is a computer whiz-kid. He got a topographical map of Manila and showed, if there was a one-metre rise in sea-levels from global warming, how many parishes would be gone. At three metres, the whole diocese was gone. Suddenly climate change became very spiritual." It was a small victory which ebbed away – because the predominant view is that climate change isn't *that* important. "I get frustrated that we're not treating it with the sort of energy as if it was a war effort and of complete focus. That's what's going to be necessary."

The Catholic Church, though, has undergone a change. Under Pope Benedict XVI, the Vatican has installed solar panels. The Pope has also suggested that climate change and an abuse of the environment is against God's will. At a Vatican conference on climate change in 2007, he called on politicians and scientists to "respect Creation" while "focusing on the needs of sustainable development". This message was mirrored by Cardinal Martino, head of the Pontifical Council of Justice and Peace, who said: "For environment . . . read Creation. The mastery of man over Creation must not be despotic or senseless. Man must cultivate and safeguard God's Creation." In the same year, the Pope made the link between ecology and poverty. The world, he said outside his summer villa near Rome, was "exposed to serious risks by life choices and lifestyles that can degrade it . . . In particular, environmental degradation makes poor people's existence intolerable."

In Ireland, the Catholic development agency Trócaire adopted climate change as its Lenten campaign for 2008. It has also, in a political move, joined the campaigning group Stop Climate Chaos which aims to compel the government to do more on reducing its emissions. Sean welcomes the move: "It is great." But he's still unhappy that so much time has been lost. "All of this was clear by the mid-1980s. If the consequences were not clear, the moral reality was. If half of what was predicted was coming down the track, then it was obvious we shouldn't go there." He feels the impacts on the natural world were forgotten and, now, the consequences are coming home to haunt us. "What will two-and-a-half million Bangladeshis do if there's a one metre sea-level rise and they can't live in the delta? Where are they going to go? Is anyone thinking of them? How many million will Ireland take?"

The question of time is something which plays on Sean's mind. "We have fifteen to twenty years to get this right or we will have changed the world for each successive generation. We will have brought in an extraordinary reality – each generation thought that their lives were better off than their parents. But with what's coming down the tracks, each generation will know that their well-being is much worse than their parents'. What I often say is: 'We are a generation which has decided, economically and across the board, that we're the last generation on the planet.' That's a highly immoral way to live."

Sean yearns for the church to give clear leadership on the subject. To that end, he's quite prepared to be extremely controversial and confrontational. In his view, US President George W. Bush's refusal to act on global warming is one of "the greatest sins of omission" because "it will make life impossible for tens if not hundreds of millions". He argues that the US bishops have "moral leadership" but don't exercise it. He says they have no problem describing abortion as "unethical and sinful behaviour", but don't do so for climate change.

Controversy isn't unknown to Fr Sean McDonagh. During a recent trip to Melbourne, Australia, he argued that the Church must consider environmental concerns on every decision, even if it means reassessing contentious issues, including birth control. He told the Australian media: "For the sake of future generations, we need to lower population, alter consumption levels and promote more resource-efficient technologies. This has ramifications for the Catholic position on birth control and for the modern growth-oriented industrial model of development which has been the principal cause of ecological devastation in our world today."

While thinking globally, he is also trying to act locally. He has been the main mover behind a pastoral letter on climate change which is currently being assessed by the Bishops' Commission. It's hoped it can be published for reading in churches for the Feast of St Francis in October.

Another project is looking at the Columban Centre at Dalgan Park, County Meath, which is right beside the Hill of Tara. At the back of his mind, maybe, are the actions of the Dominicans in County Wicklow. "We have a wonderful building, but it has not been made climate-change-friendly. We farm about 420 acres – mainly dairy – but our carbon footprint is rising with methane from the cattle. So what can we do? We will probably not be here in twenty-five years from now. Is there any way this can be put in trust? That's how I see my life over the next ten to fifteen years – as long as God gives me – along with ecological education."

At the end of our conversation at Dalgan Park, in Father Sean's neat office, I ask if he still has a sense of hope for the future. "Am I optimistic? I wouldn't necessarily be. It hasn't got anything to do with me – it's got to do with what the UN's IPCC is saying. People are not looking at the consequences. I'm a celibate. But if I had children, or grandchildren, I'd be really exercised. What kind of world are we leaving for them?"

20

Forest Fire

Ian Wright believes that the suggestion that wood pellets could be a green replacement for oil and coal is nothing short of "crazy". It's a subject which really exercises him. "The pellets are shipped in from Belgium, Canada and Argentina. It's more than likely their carbon footprint is as much, if not more, than oil."

Pellets are made from sawdust, a by-product from sawmills which turn trees into timber. The reason pellets are being imported is that we're not generating enough sawdust to satisfy demand. What concerns Ian deeply is a possible move in Ireland to make pellets directly from trees. "To cut down a tree, reduce it to sawdust, dry it to two per cent moisture, squeeze it into pellet form and then deliver it around the country . . . it's not a fuel resource we should be using."

Ian's bottom line is that pellets make economic and environmental sense only if they're made from sawmill residue. The energy required to make pellets from trees would outweigh their renewable value. Importing them is nearly worse. "Pellets are a ridiculous way to go. I think we got seduced by this pellet thing because we want the solutions to be simple."

Trees are important in the context of climate change because, according to the UN's Forum on Forests, tropical deforestation accounts for about twenty per cent of human-generated CO_2 emissions. It's estimated that six million hectares of virgin forest are lost every year through logging and other human interventions. While forests act as "carbon sinks" – converting CO_2 into oxygen during a process called photosynthesis – the issue is complicated. That's because some older forests may actually be net-emitters as the large proportion of older trees begins to decompose. Yet the UN is clear: the primary driver of carbon emissions from forests is human activity, such as deforestation and forest degradation.

In Ireland we have had a different problem – most of our trees have already been chopped down. In the 1940s, only 1.5 per cent of our land was covered in trees while the European norm was around twenty-four per cent. Due to concerted efforts to plant trees in recent years, the figure stands at roughly nine per cent here. The upside of this planting is that it has the capacity to reduce our emissions while also eliminating the need to import wood for the construction industry.

Mistakes have been made though – the most obvious being the planting of trees in bogs. Peatlands are carbon sinks, holding an estimated 5,000 tonnes of carbon per hectare. In this complex natural system, although bogs accumulate carbon, they also emit carbon dioxide and methane. The most up-to-date view is that these effectively cancel each other out. However, to plant trees, you have to drain the bog and this leads to a vast increase in the amount of CO_2 released to the atmosphere. According to the Irish Peatland Conservation Council, decomposing peatlands contribute at least 3.7 million tonnes of carbon dioxide to the atmosphere every year.

Going against the grain and questioning authority are part of Ian Wright's make-up. If they weren't, he might never have become one of the country's leading advocates for growing native tree species, a strong opponent of planting in bogs and a real thorn in the side of the government and its agencies. Ian's activism started when, "at five hours' notice", he and his wife Lynn stretched themselves and bought seventy acres of land outside Skibbereen, in west Cork. "We couldn't even afford to fence it." Their plan was to plant it with native broadleaf trees, like oaks, rather than "alien" evergreens.

"Naively, we applied for a grant, but were turned down and told we could only plant Sitka spruce. We were shocked." After getting no joy with the state Forest Service, which regulates grants, he decided to take it further. "Then I sent an appeal to Brussels, as I was aware the grant was EU money." The idea of mounting an appeal came after a chat with his friend Tony Lowes, who heads Friends of the Irish Environment. However, Ian and Lynn ran the appeal themselves.

"Looking back, the appeal was a wonderfully naive document. We listed all the things you could make from native hardwoods. But it was also very thorough." Europe was less than impressed by the government's arguments and, on their first campaign, Ian and Lynn secured victory. "The State folded quite quickly and allowed us, I think, to grow sixty per cent broadleaf on that site." Rather than celebrate a big win and forget about it, the whole affair served only to deepen his involvement. "I should have been happy, but I was so incensed at the policy."

That was 1996. Today, Wright is the project coordinator of the Irish Natural Forestry Foundation. His expertise is widely recognised and he is given platforms, like *The Irish Times*, to express his views. His targets are usually either the Forest Service, which comes under the Department of Agriculture, or Coillte, which describes itself as a commercial company owning one million acres of land, most of which is forested.

He regularly writes about the issue which first got him involved in campaigning – the prioritisation being given to "soft" Sitka spruce over native "hardwood" broadleaves. "What happened was no accident. The 1996 Strategic Plan recommended planting 20,000 hectares a year of predominantly Sitka spruce on unproductive land to achieve seventeen per cent forest cover by 2035. Many plantations were on old woodland sites. Recent research has found that if conifers on ancient woodland sites are felled and replaced with more conifers, then the wildlife dependent on ancient woodlands will not survive."

Log on to the Department of Agriculture website and it talks about developing forestry in Ireland in an economic and sustainable manner which is compatible with the protection of the environment. Ian maintains that that is all fine but the commitment to Sitka spruce and other conifers is clear in the Strategic Plan. "Any hardwood policy planting programme in Ireland is premature until the current strategic policy is largely achieved."

Officials charged with managing the nation's forests have got to know a lot about Ian Wright and the INFF because he regularly takes complaints to Brussels. "The EU was there with open arms because it asks countries: 'Have you done such and such?' And they say: 'Yes.' But Brussels doesn't have European police to go around and see if it's true. They rely on organisations like us or whistle blowers."

One of their most successful weapons, however, was electronic: they established a weekly online newsletter called *Forest Network News*. "We put in every Dáil question about forestry policy and the ministerial response." With opposition parties only too happy to highlight government problems, there were plenty of questions. "Everyone in Brussels was reading this stuff a few days after it happened."

But *FNN* didn't confine itself to the Dáil and news releases, it also reported on any policy statements. "I would go up to Dublin for a meeting with the Forest Service, and afterwards ring Tony Lowes about what took place. Before I reached west Cork, it was on the computer screens in Europe. You could see the Forest Service thinking – have we said anything that's dangerous? They'd invite me to meetings but, because they weren't used to being challenged, would often shoot themselves in the foot."

Ian's primary problem is with the Sitka spruce. "It's not a native tree. Once you bring in an alien species, there's not many things that are going to live in it – like bugs, birds or critters." He also argues that it has a negative impact on the environment. "Because of our soil type, the tree also creates an acidification problem." This problem is exacerbated by planting on bogs, which are already acidic. "It was an unbelievable thing to be doing and the fact that they are still doing it is just crazy." He contends the Sitka are also inherently weak. "The trees are not designed for our climate and so they blow down."

Sitka has got such a bad name through the large plantations which have covered the country that it's easy to forget about the tree itself. It's the largest spruce in the world and the third largest conifer. Native to the west coast of North America, the tree can live up to 700 years. However, as a tree which can be used for timber production and in the making of musical instruments, it is usually felled after forty years or so.

The government and its officials will tell you that Sitka is a fast-growing crop which can be harvested in thirty-five years while an oak takes a hundred. This means that, as they grow quickly, they provide a better financial return. Given that a hundred years ago Ireland had only a tiny amount of forest cover, it's logical to have a fast-growing crop which can stem the requirement for importing wood. On the issue of the environment, they point out that regulations are in place to ensure that hotspots like Wicklow and Kerry are managed to avoid acidification. Ultimately, the trees provide timber, which means we increase self-sufficiency and lower imports.

How the trees are cut down is another major fault-line between the two sides. The government policy is that up to twenty-five hectares of forest can be clear-felled – cut down in one go. Ian argues there should be "constant forest cover" in which the forest remains but trees are felled strategically. "We have not planted any forests here – they're plantation crops. We're kidding ourselves if we say we've now got ten per cent of the country forested. We have two per cent, and the rest is a tree crop. It's unbelievable to look at a forest and not see it as a complete eco-system. The average clear-fell in Europe is 2.5 hectares."

Ian Wright moved from London to west Cork in the mid-1970s, attracted by the remoteness and what he calls the "anarchy of the countryside". The love affair was immediate. "I think I worked it out the moment I arrived thirty-five years ago." There were laws, but people worked their way around them. "When you went for an MOT you didn't take the car because you knew the garage owner. 'Does it start and stop? Does it have brakes?' And it's signed. I loved that." It certainly made employment easier to find. "Everything was a deal. Could I teach in the arts school without any qualifications? In England, it was just tough if you failed at the Royal College. In Ireland, 'You went to the Royal College? Failed with distinction? Sure that would be grand.'"

Ian and his wife Lynn bought three acres outside Skibbereen and decided to "live the hippy dream". The move was prompted by reading, among other things, the 1960s bestseller *Silent Spring* by Rachael Carson, which told a story of how the environment was being harmed by pesticides and claimed that the chemical industry was covering up the effects. "We milked cows, killed pigs and got totally worn out eating rancid butter and living on fifty pence a week." After a while they bought another ten acres and decided to build their own eco-system. "We just started planting trees and digging lakes and ponds and saying: 'Let's make this place for bugs and critters.'"

Their pilot study, which was Lynn's brainchild, really took off. "It's so rewarding to go from a place where it was rare seeing a butterfly to seeing seventeen different species in one day. We've tried to get as many habitats as possible here with flower-gardens, shrubberies and have even dug a lake. We're very fortunate to have a couple of acres of bog and have fields which have not been farmed in fifty years. There are lots of species of grasses, for example. We're still trying to add to those habitats."

The project was important for Ian because it showed he was not simply a negative campaigner but someone who tries to put his ideas into practice. Given the clear hostility towards his idea of constant forest cover, he decided to prove that the development of forests could happen in a totally different way. The opportunity presented itself at Manch estate, which is less than five kilometres from the town of Dunmanway, also in west Cork.

"I was always very aware of how necessary and important it was to have a showcase. I think I was really looking for a twenty-acre site but ended up managing 320." The Manch estate is owned by the Connor family who, along with Ian and key corporate sponsors, are turning the land into a sustainable enterprise. They have already planted 120 acres with native Irish trees and while things are really moving now, it was a struggle in the beginning. "What I wanted to be able to say was, 'These are the same grants everyone else is getting, this is what you should be able to do.'"

The problem was getting access to grants to pay for the trees – the rules went against everything Ian believed in. "To get the biggest premium, you would put 6,600 beech trees – and nothing else – on a hectare. I disagree with a pure monoculture of beech, just as much as Sitka. So we said to the officials, 'We want the maximum grant, but we're not going to put the pure crop in.'" It could have ended there but, this time, it didn't. "The whole thing was an amazing example of the NGOs, plus the biggest private forestry company in Ireland and the Forest Service working together and saying: 'Let's see if we can come to common ground.'"

They worked out a system of planting which everyone was able to live with. "We came up with a plan to grow oaks in groups and surround them with something else. If these groups of oak are thirty metres apart, we can say, 'One day this is going to be an oak wood – and therefore we want the oak grant.' It just so happens that we've got cherry trees and eighteen other species surrounding them. Fair dues to the Forest Service, they gave us the oak grant." Where there was often hostility, there's now at least a form of mutual respect. "These are guys I've written about, pilloried in the press and taken cases against in Europe and they all come down here now." An indication of his influence is that Manch was officially inaugurated by the then European Commissioner Franz Fischler and Agriculture Minister Joe Walsh.

Manch is a centre of excellence, a "vehicle for dialogue" and, most recently, an educational resource. "What we've set up is fairly unique in Ireland. We've got primary and secondary school kids coming in for what's called an 'outdoor classroom' concept. Kids do maths, science and geography outside. And we're just so fortunate to have so much space." The interaction with the pupils has also been an eye-opener for Ian. "I didn't realise the power of the kids. They bring the parents along and we do open days and courses."

The Manch project has received significant funding from the Forest Service and is looking now for private investment. It's also undertaking trials on behalf of the National Council for Forest Research and Development and Teagasc. After years of mutual criticism, co-operation is now the name of the game. "The only way this thing is going to work is if you break down the barriers. You have got to get a dialogue going which, up to now, has not really been allowed."

As there's been a thawing of hostility between the NGOs and the State bodies, Ian is hoping that his European influence will begin to rub off on policy makers. "Look at Austria, which is sixty-five per cent forest. Every Council, by law, has to use all of its hedge cuttings, chip them and burn them. It's viewed as a fuel resource, even though they've got all that other wood. We're still cutting thousands of miles of hedges and leaving it there for the farmer to clean up or not clean up."

When it comes to the great biomass debate, Ian deems pellets to be a "crazy" way of moving forward but thinks burning chipped wood is a real viable option. Chips are substantially different from pellets because they don't have to undergo the energy-intensive process of being dried and then crunched into shape. There is, however, a learning curve. "We haven't really learned how to chip yet. In Ireland, we fell and leave trees for six months and then hope to chip them. In Austria it's eighteen months to two years. We've got to have a programme in place as these trees are going to have to come down. It is a massive fuel resource which we ought to use and, in the short term, could make a huge difference to Ireland. The major problem is over what we do with the land afterwards."

Where he finds policies that aren't working, you can be sure that Ian Wright will revert to being a fighter. One recent battle was over government biomass grants which were in place for anyone firing their boiler with wood pellets or wood chips – but not logs. "The reason for excluding logs was really cynical and blatant: the people who can burn logs are those who can cut the logs themselves and so, with no purchasing involved, there's no VAT accruing to the government." Partly due to a campaign by Ian and others, a twenty per cent grant for log boilers has now been introduced. Lynn acidly chips in: "They might backdate it, mightn't they?"

Ian believes that one of the reasons Ireland has had what he calls "a bad record on forestry" is a general lack of knowledge. "The biggest thing is that we don't have a tradition of forestry here. It's not part of the culture. We were down to one per cent at the turn of the last century. Then we had a state policy that we could not plant trees on good agricultural land in the 1950s." He even thinks that there was a section of the population who didn't like trees for historical reasons. "I think there was the landlord factor as well. It's why people in particular didn't like beech trees, because they were part of the British estate. The trees were not for the people."

The ambivalence towards trees, he feels, can be turned around when the public actually get the chance to see things first-hand. As well as working on the Manch project, Ian and Lynn have bought another seventy acres of land which is located about seven miles away from their home. "We did the same thing again. We planted forty acres there and created lots of habitat. It's very rewarding. We've got ten acres of wetlands and lots of lakes and ponds."

After charging around the world to exotic places like the Indian Ocean, home is the place where Ian feels he's really achieving something. "The knock-on effect of these projects has been huge. This has inspired so many other people to plant trees or create habitats. We have dug loads and loads of lakes and planted hundreds of thousands of trees for people. They just ask you: 'Can you come to my house and do the same?' That's really exciting."

One small irony is that Ian and Lynn Wright have ended up planting some Sitka spruce on their new lands.

"Trees have to have shelter. So there's actually a role for Sitka, because they do grow quickly and, once you get the shelter, you can grow anything else." Maybe it's not just the Forest Service which is changing.

21

The Gas Man

"It's very important that you don't go from denial to despair in one jump. There is some ground in between."

John Gormley, Ireland's first ever Green Party Environment Minister, is trying to explain the nimble high-wire act which is now his day job: informing the public about the dangers of climate change without scaring them off. It's a tricky task. How can you give people the information and ensure there's a positive reaction rather than leaving them depressed and de-motivated? The answer, it seems, is to articulate the negative but emphasise the positive. "While we face an uphill battle, and it's a monumental task, something can be done."

After ten long years on the opposition benches, John Gormley took control at the Custom House, headquarters of his Department, in June 2007. A few weeks later, he became the leader of his party. He's loving it. "It's been very interesting. It's been exhilarating and, at times, frustrating, but I know that it's the right place to be." It's the right place because he has his hands, finally, on the levers of power.

After years of making consistent demands that ministers take immediate action on any and every matter, how much power does the outsider find he actually has, now that he's on the inside? He pauses before answering: "I'll put it like this: I'm a lot more powerful than I was in opposition. Having said that, there are all sorts of limitations." The problem is that while he wants to move "very quickly", he has to "negotiate a series of obstacles". These include getting the civil servants onboard, then ensuring all legal issues are covered, engaging with vested interests and, finally, negotiating and securing the support of his cabinet colleagues. "So the answer is, in a nutshell: much more powerful than in opposition, but not as powerful as I'd like to be."

It's something of an irony that he is the Minister for the Environment at a time when Ireland has one of the worst records in the EU for tackling greenhouse gas emissions. As it stands, Ireland and Luxembourg are at the bottom of the league table. Accordingly, he adopts a direct tactic. "At any of the meetings, I put my hands up immediately and say: 'We've an appalling record and we're trying to sort it out.' I always believe in just going out and giving the warts-and-all picture." As happens throughout our interview, he then balances his comments with a positive. "I believe that in Ireland, because we're a small centralised country, and there's a cohesion others don't have, it is possible to get things moving when we put our minds to it."

John Gormley's first foray into politics was the 1989 General Election. His first success came two years later when he was elected to Dublin City Council, becoming Lord Mayor in 1994. When he finally won a Dáil seat, in the Dublin South East constituency, it was in dramatic circumstances. John Gormley was the last TD to be declared because he was separated from the PD's Michael McDowell by just a handful of votes. I was RTÉ's reporter at the count, which dragged on for a week as lawyers got involved and the nation became engrossed. At times it was bizarre – at one point the returning officer and party officials spent hours using magnifying glasses to assess if individual ballot papers had perforations from the polling-booth punchers and, therefore, were valid. In the end, he defeated McDowell by just twenty-seven votes. However, their war had only just begun.

Between the time he was deemed elected in June 1997, and the time he became Minister in June 2007, John Gormley was an outspoken critic of Fianna Fáil-led governments. In the run-up to the last general election, he launched a blistering attack on then Taoiseach Bertie Ahern's financial record. "On Planet Bertie you can get loans from people – that you don't have to pay back. On Planet Bertie you can save €50,000 – without a bank account. And on Planet Bertie, climate change doesn't exist. All that stuff is made up by Trevor Sargent." He also got stuck into Fianna Fáil. "There's a strange cult called Fianna Fáil, a type of religion without vision or values; and every year in August they go on their annual pilgrimage to one of their sacred sites, the tent at the Galway races."

It was a virulent assault which concluded in emphatic terms. "It's a planet that we Greens would like to avoid. For let there be no doubt, we want Fianna Fáil and the PDs out of government." A few months later, the Greens had joined them in government. The stated reason why the party went into coalition with Ahern and Fianna Fáil, despite everything, is climate change. "Without being alarmist, all the science shows us that we have a ten-year opportunity to tackle the problem. That's it. I simply couldn't afford to be sitting on opposition benches for another five years."

Given that volte-face, something the current opposition has been constantly sniggering at, it's politically imperative for Minister Gormley to be seen by the electorate to deliver on global warming promises. His over-riding priority is what he calls "putting a price on carbon". And it's something which has to happen now. "I'd like to see the introduction, in 2008, of a carbon levy." This is essentially a tax on fuels which contribute to global warming, such as petrol, in the hope it will dissuade people from using them. "If we don't get it in this year, it'll be very hard to bring it in at any stage. It would be a fairly major shift in policy." The proposal is currently before a Commission on Taxation. If the Minister is successful in getting it into the 2008 budget, it would be a singular victory, as the measure has been part of the National Climate Change Strategy since 2000 but has never been implemented.

A second key priority is to reduce Ireland's greenhouse gas emissions, which are currently nearly twice as high as specified by the Kyoto Protocol. It's something John Gormley says will only happen if all the cabinet gets involved. "We do have to bring them all together. We need every single department to step up to the plate." Interestingly, by accident or design, he goes on to single out only one department, which also happens to be the sector with the largest emissions. "We need agriculture to come forward, taking good advice from the Environmental Protection Agency, and say: 'This is how we're going to make our cuts.' And every other department to follow, saying: 'These are our cuts and this is how we're going to do it.'"

But the Minister is not in charge of these other departments. Within his own, he can make things happen. He has changed the motor tax so a vehicle's emissions, rather than the size of the engine, determines the bill. And he's tackling the planning laws. "We have to stop bad planning. This will ensure the new places that we build have access to schools and amenities." New developments should be connected to public transport, so people don't have to rely on cars. "That's the biggest uphill battle we face in the country. The latest statistics that I have show our emissions from the transport sector will have increased, if unchecked, by 256 per cent between 1990 and 2020. It's really, really difficult. We're looking for emissions cuts, but we're looking at this trajectory."

While planning for the future, John Gormley has had some of the previous government's decisions come back and bite him. On the night he received his ministerial seal of office from President McAleese, RTÉ News reported that his predecessor, Dick Roche, in one of his last acts before vacating the Custom House, signed the order which allowed for the destruction of a national monument near

Tara, to make way for the M3 motorway. Just a few months later, Dublin City Council secured planning permission and licensing for its massive waste incinerator in Gormley's constituency. In both cases, the opposition parties charged that he could have reversed the decisions but was failing to live up to his promises.

John Gormley's response is usually to come out all guns blazing. It's a characteristic which served him well during the last election when he decided to tackle his old sparring partner Michael McDowell, then Minister for Justice and leader of the Progressive Democrats. Gormley cycled alone to a PD media event in Ranelagh. There McDowell was planning to climb a pole to launch a new party poster, repeating a successful similar stunt in 2002. This time it fell apart as Gormley strode into the middle of it waving a PD election pamphlet, shouting : "You say that we're going to raise corporation tax. That is a lie. Withdraw that." McDowell was clearly taken aback but responded by accusing the very unwelcome guest of having "lost it". But John Gormley was tenacious and refused to back off. "I'm not taking any nonsense from you any more." When the count in Dublin South East was completed less than two weeks later, he beat McDowell for the final seat by just over 300 votes.

When not fighting rearguard action, the Environment Minister is making plans. One major objective is to strive to get Irish houses to the point where they have zero-carbon emissions. "You should be able to turn on your computer and heat your house using electricity generated by renewable sources." Energy-guzzling devices like the tumble-drier wouldn't be needed. "You would be able to dry your clothes because the air in the house would circulate in a particular direction." He believes the public will go for it because energy bills would disappear at a time oil prices are high. "It's a new concept but I think it's one that's easy to grasp." The target date for standardising such housing was 2016 but he feels that should be brought forward. He is also focusing on what to do about the existing houses. According to his department, thirty-five per cent of the entire stock has been built over the past ten years. In the vast majority of cases, this was to low energy efficiency standards when compared to our European neighbours. "We will have to go back and decide how we're going to fix the things we did so badly. We're going to have to retro-fit thousands upon thousands of houses – millions if the truth be known – and insulate them properly."

John Gormley contends that making the house of tomorrow a reality, in just a few years, isn't pie-in-the-sky. "The technology is changing all the time. The building materials are changing. A zero-carbon house is currently possible." He's steadfast in his belief that it can be done. "A lot of the solutions I'm talking about are not high-tech. They're just about using your common sense and designing your house in such a way. To me, it does not make sense at all to be turning on your immersion and all of the heat just going through the walls of your house. What is so high-tech about a zero-carbon house? Well, very little really. It's just a new way of thinking." The construction industry has already expressed the view that the Minister is moving too fast. For him, things are not moving fast enough. "I think that we have to go there quickly – more quickly than anyone has anticipated. The thing about climate change is that no sooner do we have statistics published than they are out of date and we're bringing forward dates on when ice will melt."

Government's slow pace of change is an ongoing source of frustration. "I brought it up yesterday with managers of the local authorities. I said, 'I want you to come forward and tell me how you are going to cut emissions.' I've put energy efficiency and CO_2 cuts into how the councils' performances are rated by the government. But they are slow to act." He also takes a swipe at his own department, pointing to spotlight bulbs above his head – which he leaves off. "Those are bloody terrible energy wasters. It's the simple things: Why is the heat on all the time? Why can't you change bulbs? What's the problem? It's intensely frustrating for me, although six months is a short enough period of time to be in office."

The issue of bulbs turned out to be a major controversy, even though it all started so well. In December 2007, the Environment Minister announced that Ireland would become the first country in the world to ban the traditional, but energy inefficient, light bulbs by January 2009. He maintained that the switch to new long-life low-energy bulbs would save €185 million in electricity costs while also saving "700,000 tonnes of carbon dioxide emissions every year". He admitted that there would be teething problems but they cropped up quicker than he expected. The bulb company Solas made it known that the Minister hadn't taken their concerns on board. The opposition claimed that the European Commission would oppose the measure. Chat-rooms started to call him "40-watt-

Gormley", accusing him of executing another U-turn. Questions were asked, and not answered particularly well, about plans for the disposal of the new bulbs, which contain mercury. However, he did also receive support. "Greenpeace hopes that Ireland's decision will light the way for the EU and the rest of the world." By January, the department said they had received assurances from a Commissioner that the ban would be allowed to come into effect.

The slow pace of change in Ireland is also reflected on the international scene. John Gormley represented Ireland at the big United Nations Environmental Summit in Bali, Indonesia, in December 2007. He describes progress there as "tedious, slow and lacking urgency". The aim of the talks process, involving nearly every nation on the planet, is to agree a successor by the end of 2009 to the Kyoto Protocol. While the deal in Japan compelled developing countries to introduce a cap on their emissions, the next deal is aimed at getting all countries to reduce their CO_2 output.

Given the scale of the discussions, the Minister isn't surprised that things progressed at a snail's pace. "I knew it was going to be difficult to get consensus among so many countries." The EU delegation was led by the Portuguese, who consulted with ministers like John Gormley every morning. "We would give our input and make our observations on the talks process. When they'd go back into the plenary sessions, we'd then hold bi-lateral meetings with ministers from other countries or meet environmental campaigners. It was very productive as you get new ideas." Inevitably at these meetings questions are asked about whether it's absolutely necessary for thousands of delegates to participate in the talks process. "You do have to ask if we should all be going to these big conferences and flying on airplanes. It's a difficult one. I was meant to fly to Brussels today for a meeting. I said I wouldn't bother as I just didn't feel it was necessary."

While he was not negotiating on behalf of the EU, the Minister was certainly putting in the work. Indonesia is eight hours ahead of Ireland. At one stage during the summit, I was giving a live update on RTÉ's nine o'clock TV news – it was five in the morning in Bali. Shortly afterwards, I spotted John Gormley on his way back into the talks. He must have only had a few hours sleep as he'd been up late as well. A few days earlier, he'd travelled with the media to a beach where turtles were dying as a result of stronger waves and warmer waters. Driving back to the conference centre, he slumped forward in the passenger seat. Three o'clock in the afternoon and he was sound asleep.

The key development from Bali was that the United States committed itself to participating in the negotiations on a Kyoto Protocol Mark 2. It does not necessarily mean they'll ultimately sign up, but John Gormley is convinced that there will be a sea-change when the US gets a new President. "I believe the US is going to take its rightful place and lead the debate. In a sense it's already happening there at city and state level."

However, he does not underestimate the global problems that remain, particularly the growth of China which, as yet, does not have a cap on its emissions. "China is becoming the biggest global emitter – although not on a per capita basis. Their economy is expanding by about nine per cent per year. Their growth in terms of population and housing is phenomenal. They are building a coal-fired power station every four days or so. In my view it's going to become the world's biggest economy. And the US fears them." It's going to be a difficult task to negotiate a deal which will seriously tackle greenhouse gases but which all sides can sign up to.

Given his optimistic nature, it is something of a surprise when Minister Gormley speculates on what might happen if the talks fail. "My fear is that if we don't sort those difficulties out, then we are getting into a conflict situation. I mean, most wars are resource wars." How real does he believe the risk is? "I think it is real. I think we'd be very foolish to underestimate the potential for conflict. Those scenarios have been mapped out in various defence departments. We've already seen it in some parts of the world; people have said that Chad and Sudan is a climate change conflict." Because of the failure of international treaties to tackle climate change to date, does he see any hope? "You could say the United Nations is an imperfect vehicle to deal with it, but it's the only one we have right now."

The moment John Gormley realised the vulnerability of the planet, and was forever more set on an eco-path, was while reading the 1972 book, *Limits to Growth*, by the think-tank, The Club of Rome. The concept of the book, which has sold tens of millions of copies, was to use models to estimate the potentially devastating impact on the planet caused by rising world population, industrialisation, pollution, food production and resource depletion.

While the book has been criticised over the way it arrived at its conclusions, he believes that a lot of what it contained has been proven to be accurate. While concerns over climate change today are extremely troubling, the intense international focus on global warming is a cause for hope. "Climate change will be the only story in town. I've been around dealing with the environment a long time and I remember in the 1970s when there was a peak of awareness and then a decline. It came back up again in the late 1980s. But now environmental awareness is only going one direction and the trajectory is way up." Like many others, he credits the IPCC reports and Al Gore's film with making climate change mainstream. "You know it's arrived when Paris Hilton and beauty queens are talking about what they're going to do for the environment. That's it. It's arrived and it's not going to go back."

While risk awareness is welcome, John Gormley wants to be absolutely clear that it's only a first step towards what must become a radically changed world. "Awareness is all very well but we are leading unsustainable lifestyles. There's no question about that. How do we change? That's the question, because the cuts that are required are momentous. If we look at the long term to 2050, we're talking about an eighty-five per cent cut. In one sense, we're only starting out on the climate change journey."

It's not just a question of adapting our houses and changing transport and energy systems; the Minister hopes there will be a philosophical change as well. "For me it comes back to the issues of economic growth and consumerism. We don't actually require all this material wealth. Why do you have to change your car every year?

Why is bigger beautiful all the time? It's not. The things that make life worth living are very simple things. What people actually enjoy doing is going for a walk or a swim. Those are activities which don't cause huge emissions but add quality to life. It's why I think the churches have a role to play in getting people to think a different way. That would do more than anything else to cut emissions. You don't have to go to New York to do your shopping. Let's get real. If we do get real, we can continue to live as a species on the planet. If we don't, well, we've had our chips."

Epilogue

As human beings, our first response to a major crisis, like global warming, is often to stick our heads firmly in the sand. Some people find the scale of climate change to be so immense that they simply switch off. Out of sight, out of mind. For others, the easy option is to contend that the jury is still out. Given that the row over the scale of the problem and the most effective responses will rumble on for decades, it rather handily obviates the need for any personal action.

The second stage, for many, is to become either hysterical or, as Minister John Gormley phrases it, to jump from denial to despair. Sections of the media certainly have played a role in hyping the problem. Global warming is an enormous issue, but it does not mean that doomsday is inevitable. And while the days of Arctic sea-ice are numbered, I don't believe that polar bears are going to disappear any time soon.

A third and more complicated response is the rush "to do something". The drive to replace fossil fuels with bio-fuels is a good example. It simply isn't smart to cut down rainforest in order to produce a crop which can be used as a replacement for petrol. It does not make sense to take up massive proportions of agricultural land when the yield per acre is low and the poor of our world are going to starve as a result. A cool evaluation of a hot topic is clearly called for.

Currently, in what I would call the fourth phase, all responses are being subjected to a more rigorous cost-benefit analysis. What difference is Ireland going to be able to make to such an immense global problem? Should we spend €1 billion on carbon credits to ensure we "achieve" our Kyoto targets or would the money be better spent on research? Should the EU unilaterally undertake to significantly reduce its emissions by 2020 when the big five developing countries – China, India, South Africa, Mexico and Brazil – will, in all likelihood, do the opposite?

These questions are pertinent and relevant but also lead to uncertainty. This was highlighted in a survey conducted on behalf of the Department of the Environment in March 2008. From the findings, it's clear that there is a huge information gap. Nearly one-fifth of the 1,000 people polled felt that climate change was mainly a natural change, rather than a problem caused by human behaviour. Seventeen per cent said the topic never came up in conversation with their friends.

Minister Gormley will be encouraged that ninety-three per cent of those questioned felt that every country, no matter how small, has its part to play in tackling global warming. Ninety-six per cent of those polled wanted the government to take an active role but only thirty-four per cent thought that currently was the case.

Things become less clear when it comes to taking action. For example, the survey found that eighty-one per cent of people agreed with government proposals to introduce carbon taxes which would increase prices on those fuels that are more damaging to the environment. But when it came to paying, opinions seemed to change. For example, forty-three per cent opposed an eight per cent increase on a litre of car fuel; fifty-six per cent opposed an eight per cent increase on a delivery of home heating oil; fifty-four per cent opposed a six per cent increase in a typical gas bill; fifty-one per cent opposed a ten per cent increase on a bag of coal; while fifty-four per cent opposed a ten per cent increase on a bale of peat briquettes. When it came to encouraging water conservation through new household charges, fifty per cent said they were against it.

Given this public mood, and the uncertainty over how to move forward, there is, in my view, a substantial possibility that people will focus solely on the evaluation of the solutions and forget about the problem. In essence, it lends itself towards reticence about taking any action at all. If we are not entirely sure that a solution is going to work, then maybe it isn't worth the expense? If the developing countries are not prepared to take significant cuts in their emissions, is there any purpose in the EU taking action? Given that the solutions are so expensive to investigate and difficult to nail down, should we not just plough on and let the next generation sort it out?

This is a real danger. Why? Because it is overly focusing on the concerns of today rather than the consequences of inaction on tomorrow. On so many occasions, we hear the lofty ideal expressed that people are doing things today to improve the lot of our children tomorrow. Well, if you ever wanted an acid-test of whether this ideal works in practice, global warming is it. Even if the scientists are only partially correct, then climate change is going to be the biggest issue for this, the next and subsequent generations. It is, in my view at least, beyond debate that the actions we take today will have monumental impacts in fifty years' time – for good or bad. Failure or refusal to act could well mean that this generation will be cursed by those who follow us because, while we may not be in command of all the facts, we have a pretty good idea what the likely consequences are going to be. Grim.

Paul Cunningham
Dublin, April 2008

Leabharlanna Poibli Chathair Bhaile Átha Cliath
Dublin City Public Libraries

In memory of my friend, John O'Briain,
from Sandyford, County Dublin,
who died suddenly in March 2008

Direct to your home!

If you enjoyed this book why not
visit our website:

www.poolbeg.com

and get another book delivered straight to
your home or to a friend's home!

www.poolbeg.com

All orders are despatched within 24 hours.